Business Math Essentials

Robert L. Dansby

Columbus Technical Institute

Prentice Hall

Upper Saddle River, New Jersey   *Columbus, Ohio*

Library of Congress Cataloging-in-Publication Data

Dansby, Robert L.
  Business math essentials / Robert L. Dansby.
      p.   cm.
  Includes index.
  ISBN 0-13-079978-5
  1. Business mathematics.   I. Title.
  HF5691.D363   2000
  650'.01'513—dc21   99-25814
                                                      CIP

Cover art: © Superstock, Inc.
Editor: Stephen Helba
Production Editor: Louise N. Sette
Production Supervision: Carlisle Publishers Services
Design Coordinator: Karrie Converse-Jones
Text Designer: Laurie Janssen
Cover Designer: Tanya Burgess
Production Manager: Deidra M. Schwartz
Marketing Manager: Shannon Simonsen

This book was set in Palatino by Carlisle Communications, Ltd., and was printed and bound by Banta Company. The cover was printed by Banta Company.

Printed in the United States of America

10 9 8 7 6 5 4 3 2 1

ISBN: 0-13-079978-5

Prentice-Hall International (UK) Limited, *London*
Prentice-Hall of Australia Pty. Limited, *Sydney*
Prentice-Hall of Canada, Inc., *Toronto*
Prentice-Hall Hispanoamericana, S. A., *Mexico*
Prentice-Hall of India Private Limited, *New Delhi*
Prentice-Hall of Japan, Inc., *Tokyo*
Prentice-Hall (Singapore) Pte. Ltd., *Singapore*
Editora Prentice-Hall do Brasil, Ltda., *Rio de Janeiro*

Preface

Much has been written in recent years about the declining ability of Americans to perform basic math tasks. Much of this comes not from a lack of ability, but from a fear of math that students develop at an early age. *Business Math Essentials* is designed to give students confidence in their math ability. It is written in a friendly, nonthreatening dialogue that will not overwhelm students. It always talks *to* students, never down to them.

We live in a fast-moving, rapidly changing world. Skills in basic business mathematics are needed to enable today's workers to obtain a position and to advance in that position. According to the latest research, the average professional (bookkeeper, accountant, office worker, manager, and so on) will change positions at least six times during a working career. Many of these changes will take place because individuals will have the educational and occupational skills needed to keep moving up the success ladder. *Business Math Essentials* is designed to provide the basic math skills needed to enter the workforce. It is a user-friendly text with the following objectives:

1. Thoroughly reviewing the fundamental operations of arithmetic: whole numbers (addition, subtraction, multiplication, and division), fractions, decimals, and percents.
2. Developing the student's ability to apply the basic math skills to common business situations.
3. Developing the student's ability to solve common business problems involving discounts, payroll, interest, markup, depreciation, inventory, and banking.
4. Developing the student's ability to use shortcuts and work mentally with speed and accuracy.

Business Math Essentials can be used in remedial math courses because of its basic and step-by-step approach. Yet its comprehensive coverage of major business math topics will also allow it to be used in traditional business math courses. The text has been designed to allow students to work on their own or receive instruction in the traditional classroom setting. It teaches by introducing concepts, supported by examples and immediate feedback. Text matter and problems move from easy to more challenging.

ACKNOWLEDGMENTS

I would like to express my sincere thanks to my academic colleagues who offered many excellent suggestions and recommendations during the review process of this book and its supplements. This project would not have been possible without their expert assistance. Their time, experience, and expertise are greatly appreciated: Richard P. Burgholzer, Rochester Business Institute; Janet Caruso, Briarcliffe College; William Kamenoff, Baltimore City Community College; and Erwin L. Zweifel, Madison Area Technical College.

Special thanks are also in order to my good friends at Prentice Hall, who are responsible for this book. My sincere appreciation goes to Steven Helba, Acquisitions Editor; Nancy Kesterson, Editorial Assistant; Louise N. Sette, Production Editor; Karrie Converse-Jones, Design Coordinator; Tanya Burgess, Cover Designer; Deidra M. Schwartz, Production Manager; and Shannon Simonsen, Marketing Manager.

Accolades and warm regards are in order for my Project Editor at Carlisle Publishers Services, Kate Scheinman, for her enthusiasm, promptness, constant availability, and unflagging devotion to the project. I would also like to thank the Carlisle text designer Laurie Janssen and copy editor Elaine Honig.

Finally, I wish to thank my family for their love and support: wife, Barbara; son, Robert (Champ), and daughter, Allison (Alli).

Contents

v

Contents

This book is dedicated to my children:
To my little daughter, Alli; an unexpected surprise and an indescribable joy. And to my remarkable son, Champ, who in the face of an incurable disease showed courage and resolve much beyond his years. He definitely has the right nickname.

section

I

Whole Numbers: Addition and Subtraction

unit 1.1

Reading and Writing Whole Numbers

After completing Unit 1.1, you will be able to:

1. Define whole numbers.
2. Read and write whole numbers.
3. Identify place values.

A **whole number** is a number that we use for counting and numbering. We use the following ten symbols, called **digits,** to represent whole numbers:

0, 1, 2, 3, 4, 5, 6, 7, 8, and 9

Using these ten digits, we can express any number, large or small. Numbers above the ten basic digits are formed by using combinations of the ten basic digits. The number "10," for example, is formed by using a combination of the first two digits.

THE PLACE VALUE SYSTEM

We can determine the value of each digit in a number by looking at its **place,** or its location, in the number. Thus, each digit in a number has a **place value.** To illustrate place value, let's look at the following chart, which shows the number "5" in three different *places.*

Billions			Millions			Thousands			Hundreds			
Hundreds	Tens	Ones	Hundreds	Tens	Ones	Hundreds	Tens	Ones	Hundreds	Tens	Ones	
											5	Example A
										5	0	Example B
									5	0	0	Example C

The first 5 (Example A) is in the *ones* place; therefore, it has a value of 5 ones. The second 5 (Example B) is in the *tens* place. Thus, it has a value of 5 tens and no ones, and is written as *50*. The third 5 (Example C) is in the *hundreds place; thus, it has a value of 5 hundreds, no tens, and no ones, and is written as 500.* Notice that each place has a value that is worth ten times the value of the place to its right. A digit in the tens place is thus worth ten times the same digit in the ones place, and a digit in the hundreds place is worth ten times the same digit in the tens place.

In the place value system, we can write numbers in a very compact way. For example, the number "256" is actually:

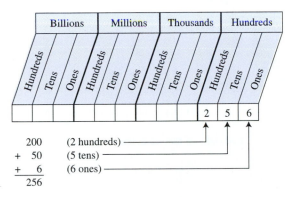

								200	(2 hundreds)
+	50	(5 tens)
+	6	(6 ones)
256

The number "256" is read as **two hundred fifty-six** and is the sum of its place values.

Self-Check

In the number 845, what is the place value of the:

(a) 8
(b) 5

Solution

(a) hundreds
(b) ones

READING LARGER NUMBERS

Using the place value system makes it much easier to express and read large numbers. For example, let's look at the following chart, which shows the number 1,216,566 with its place values labeled.

Billions			Millions			Thousands			Hundreds		
Hundreds	Tens	Ones	Hundreds	Tens	Ones	Hundreds	Tens	Ones	Hundreds	Tens	Ones
					1	2	1	6	5	6	6

We read this number as **one million, two hundred sixteen thousand, five hundred sixty-six.**

To make the number easier to read, we use commas to separate the digits into groups of three. Each group of three digits (except the ones group) is read by saying the three-digit number followed by the name of the group to which it belongs. Let's look at some other examples.

Example 1

Read the number 12,318,412 out loud.

Solution

Billions			Millions			Thousands			Hundreds		
Hundreds	Tens	Ones	Hundreds	Tens	Ones	Hundreds	Tens	Ones	Hundreds	Tens	Ones
			1	2	3	1	8	4	1	2	

Twelve million, three hundred eighteen thousand, four hundred twelve.

Example 2

Read the number 10,615,258,016 out loud.

Solution

Billions			Millions			Thousands			Hundreds		
Hundreds	Tens	Ones	Hundreds	Tens	Ones	Hundreds	Tens	Ones	Hundreds	Tens	Ones
1	0	6	1	5	2	5	8	0	1	6	

Ten billion, six hundred fifteen million, two hundred fifty-eight thousand, sixteen.

Example 3

Read the number 210,344,865,008 out loud.

Solution

Billions			Millions			Thousands			Hundreds		
Hundreds	Tens	Ones	Hundreds	Tens	Ones	Hundreds	Tens	Ones	Hundreds	Tens	Ones
2	1	0	3	4	4	8	6	5	0	0	8

Two hundred ten billion, three hundred forty-four million, eight hundred sixty-five thousand, eight.

In reading the number in the last example, notice that zeros are not read. This is because a zero means "no value." Thus, the number 500 means no ones, no tens, and 5 hundreds. In the same way, the number 3,008 means: 8 ones, no tens, no hundreds, and 3 thousands.

Also, notice that we did not use the word "and" when reading any of the whole numbers in our examples. To take another example, the number "375" is read as "three hundred seventy-five," *not* as "three hundred and seventy-five." You only use "and" to show that a whole number also has a partial unit; we will talk about this in a later unit.

Self-Check

Read the following numbers out loud:

(a) 972
(b) 12,685

Solution

(a) nine hundred seventy-two
(b) twelve thousand, six hundred eighty-five

≋ EXPRESSING WHOLE NUMBERS AS DOLLARS

To express a number in dollars, you write the dollar sign to the left of the number. Thus, "fifty dollars" is written as $50. To take another example, "four thousand, five hundred forty dollars" is written as $4,540.

Exercises

Name _____ Date_____

Write the place value of the underlined digit:

1. 5<u>6</u>6

2. 28<u>7</u>

3. <u>8</u>92

4. 1,<u>9</u>75

5. <u>4</u>,587

6. 2<u>3</u>,498

7. <u>4</u>4,593

8. 1<u>2</u>5,986

9. <u>3</u>45,687

10. 15,67<u>7</u>,890

ANSWERS

1. _____
2. _____
3. _____
4. _____
5. _____
6. _____
7. _____
8. _____
9. _____
10. _____

Write these numbers:

| | Billions | | | Millions | | | Thousands | | | Hundreds | | |
	Hundreds	Tens	Ones	Hundreds	Tens	Ones	Hundreds	Tens	Ones	Hundreds	Tens	Ones
11.								2	2	8	1	5
12.				3	8	0	1	6	7	7	2
13.			5	6	7	9	4	3	8	2	9	9
14.	5	4	5	8	9	6	3	7	5	0	0	0
15.				1	0	0	3	1	8	0	1	7
16.							9	9	0	0	0	0

Write these numbers in words:

17. 76 _____

18. 235 _____

19. 2,398 _____

20. 12,398 _____

21. 123,498 _____

22. 1,284,598 _____

23. 34,598,165   _____

24. 125,385,387   _____

25. 33,325,385,398   _____

Write these numbers in digits:

26. seventy-five

27. forty-eight

28. Two hundred ninety-nine

29. Eight hundred four

30. Fifteen thousand, three hundred three

31. One hundred two thousand, thirty-nine

32. Six million, four hundred two thousand

26. _____

27. _____

28. _____

29. _____

30. _____

31. _____

32. _____

Work Space

unit 1.2

Adding Whole Numbers

After completing Unit 1.2, you will be able to:

1. Identify the parts to an addition problem.
2. Add two or more whole numbers.

Addition is an arithmetic operation in which two or more numbers (called **addends**) are combined into a single larger number, called the **sum** or the **total**. You show addition by writing the plus symbol ("+") next to the addends. Let's look at a simple addition problem to review these terms:

$$
\begin{array}{r}
9 \longleftarrow \text{addend} \\
+\,6 \longleftarrow \text{addend} \\
\hline
15 \longleftarrow \text{sum or total}
\end{array}
$$

To add a larger whole number, you must align the numbers to be added (addends) into columns with ones digits over ones digits, tens digits over tens digits, hundreds digits over hundreds digits, and so on. Let's look at an example.

Example 1
Find the sum of 24 and 18.

Solution

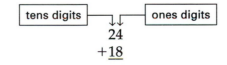

$$
\begin{array}{r}
24 \\
+\,18 \\
\hline
\end{array}
$$

The "4" in the number 24 means 4 ones; the "8" in the number 18 means 8 ones. So, these digits must be aligned before you can start adding. The "2" in the number 24 means 2 tens; the "1" in the number 18 means 1 ten. So, these digits must be aligned.

Now start the addition by adding the digits in the ones place (4 ones + 8 ones = 12). Write down the "2" under the ones column and carry the "1" to the top of the tens column, then add the digits in the tens column:

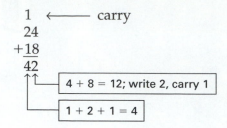

Since the sum of the tens column, 4, is less than 10, there is no carry. Let's look at another example.

Example 2

Find the sum of 1,345 and 928.

Solution

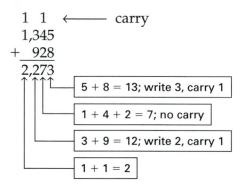

≋ CHECKING FOR ACCURACY

The accuracy (or correctness) of any arithmetic operation should always be checked. An easy way to check the accuracy of an addition problem is simply to add backward. That is, you add downward to get your sum, then you add upward to check your answer.

Example 3

Add and check: 348 + 295 + 125.

Solution

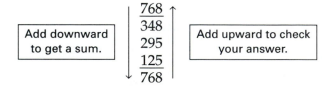

Another way to check the accuracy in an addition problem is to switch the order of the addends in a column and to re-add the column. You will get the same answer if your work is correct because the order of the addends does not change the sum.

COMBINATIONS THAT TOTAL TEN

Even in our modern age of credit card size calculators and brief case size computers, there is a great need for the ability to add numbers manually with speed and accuracy. Often you may find that you need to add numbers quickly and a calculator or computer may not be handy. And, many situations in our fast-moving business world may call for you to add numbers quickly in your head.

You can gain speed and accuracy in addition more quickly if you learn to recognize combinations of *two* or numbers that total 10. Here are some combinations of numbers that add up to 10:

$$
\begin{array}{cccccc}
 & 1 & 2 & 3 & 4 & 5 \\
+ & 9 & 8 & 7 & 6 & 5 \\
\hline
 & 10 & 10 & 10 & 10 & 10
\end{array}
$$

$$
\begin{array}{cccccccc}
 & 1 & 1 & 1 & 1 & 2 & 3 & 4 \\
 & 1 & 2 & 3 & 4 & 3 & 3 & 4 \\
+ & 8 & 7 & 6 & 5 & 5 & 4 & 2 \\
\hline
 & 10 & 10 & 10 & 10 & 10 & 10 & 10
\end{array}
$$

You should learn these combinations so well that when seeing them, you will think 10 immediately without having to add mentally. Notice how quickly we can add the following numbers using combinations that add up to 10:

5 5 > 10	3 7 > 10	8 2 > 10	
6 4 > 10	6 —— 6	6	
7 3 > 10	5 5 > 10	3 —— 3	
	2 —— 2	3 —— 3	
		4 —— 10	
30	**28**	**26**	

Larger numbers can also be added more easily if you look for combinations that total 10:

Example
Add: 853 + 657 + 486 + 324

Solution

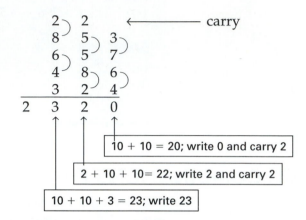

			← carry
2	2		
8	5	3	
6	5	7	
4	8	6	
3	2	4	
2	3	2	0

10 + 10 = 20; write 0 and carry 2

2 + 10 + 10 = 22; write 2 and carry 2

10 + 10 + 3 = 23; write 23

Self-Check

Add the following:

852 + 248 + 459 + 771

Solution

8	5	2
2	4	8
4	5	9
7	7	1
2,3	3	0

Name _____ **Date** _____

Exercises

Add these numbers:

1. 45	**2.** 45	**3.** 47	**4.** 86	**5.** 56	**6.** 98
15	23	18	27	38	19

7. 86	**8.** 35	**9.** 87	**10.** 75	**11.** 67	**12.** 76
41	15	78	27	39	35

13. 456	**14.** 287	**15.** 387	**16.** 186	**17.** 823	**18.** 487
34	9	126	167	638	187

19. 898	**20.** 509	**21.** 678	**22.** 386	**23.** 487	**24.** 896
198	387	287	492	224	892

1. _____
2. _____
3. _____
4. _____
5. _____
6. _____
7. _____
8. _____
9. _____
10. _____
11. _____
12. _____
13. _____
14. _____
15. _____
16. _____
17. _____
18. _____
19. _____
20. _____
21. _____
22. _____
23. _____
24. _____

25. 3,598	26. 9,076	27. 5,398	28. 8,439	29. 3,365
808	988	3,455	4,986	2,974

30. 8,794	31. 5,388	32. 4,588	33. 5,299	34. 8,766
5,986	4,694	2,499	3,268	4,563

35. 12,518	36. 35,877	37. 89,764	38. 65,747	39. 32,897
3,765	2,008	12,786	54,865	18,649

40. 345,611	41. 347,365	42. 946,127
55,833	134,587	577,387

Find the sum and check for accuracy in each of these problems:

43. $234 + 129 + 19 =$

44. $172 + 12 + 8 =$

45. $976 + 874 + 912 =$

46. $162 + 478 + 211 =$

47. $125 + 644 + 901 =$

48. $617 + 911 =$

49. $4,597 + 2,345 + 876 + 77 =$

ANSWERS

25. _____
26. _____
27. _____
28. _____
29. _____
30. _____
31. _____
32. _____
33. _____
34. _____
35. _____
36. _____
37. _____
38. _____
39. _____
40. _____
41. _____
42. _____
43. _____
44. _____
45. _____
46. _____
47. _____
48. _____
49. _____

50. 2,488 + 1,276 + 1,145 + 221 =

50. _____
51. _____
52. _____
53. _____
54. _____
55. _____
56. _____

51. 8,455 + 4,356 + 1,450 + 865 =

52. 7,745 + 2,938 + 2,386 + 75 =

53. 5,588 + 3,455 + 3,599 + 255 =

54. 9,348 + 7,459 + 2,499 + 1,235 =

55. 8,677 + 3,587 + 2,699 + 2,349

56. 3,582 + 77 + 8 + 5 =

57. $3,485 + 275 + 211 + 11 + 9 =$

58. $12,455 + 2,455 + 918 + 212 =$

59. $125,678 + 109,489 + 4,566 =$

60. $345,689 + 234,587 + 99,877 =$

Add these numbers and check for accuracy:

61.	**62.**	**63.**	**64.**
25,890	38,925	45,611	68,957
12,911	22,344	21,549	12,417
8,452	7,894	3,578	932
6,784	2,375	3,247	222
657	946	76	17

65.	**66.**	**67.**	**68.**
34,599	45,587	89,476	124,599
24,563	34,294	45,934	104,598
3,459	22,345	56,398	99,458
2,388	11,855	12,345	12,765

69. 1,245,918
1,009,213

70. 3,455,398
   999,347

71. 28,975
12,135
27,774
83,336
14,638
  6,472

72. 44,652
26,458
83,921
67,189
  5,433
  5,678

73. 47,653
13,477
11,737
99,373
  3,999
  7,101

Work Space

unit 1.2

Applications

In each of these application problems, look for words that give you some clue that you have to use addition. These are words such as sum, total, combined, and altogether.

ANSWERS

1. _____

2. _____

3. _____

1. Wayne Burke bought these books for his classes:

    Biology $20
    Math 12
    English 18
    History 28

    What was the total cost of his books?

2. Susan Kaye made these bank deposits last month: $125, $275, $321, and $508. What was the total amount that she deposited during the month?

3. Pete Leo bought a house for $38,800 and made these repairs:

    Paint $400
    Wallpaper 390
    Carpet 1,950
    Storm windows 1,250
    Tile 850

    What was the final cost of his house?

4. Spencer Variety Store made these amounts last week: Monday, $12,233; Tuesday, $9,455; Wednesday, $12,478; Thursday, $14,569; Friday, $17,450; Saturday, $23,455. What were the total sales for the week?

4. _____

5. _____

6. _____

7. _____

5. Susan Chun teaches business classes. This term she is teaching these subjects:

Subject	Number of Students
Business Math	35
Bookkeeping	25
Typing	40
Business English	15

What is the total number of students she is teaching?

6. On July 1, Dave Rodriguez bought 125 acres of land. On August 17, Dave bought another 500 acres. And, on September 15, he bought 340 more acres. How many acres did Dave purchase altogether?

7. The Simmons Company owns property that is worth $125,800; the Dyers Company owns property that is worth $250,000. What would be the total value of the properties if the two companies were formed into one company?

8. Sue Lawson earns $900 a month and lives with her parents. She is thinking about moving into an apartment. She figures that she will have these expenses:

Rent $300
Food 200
Utilities 100
Clothes 50
Car payment 225

Is Sue earning enough to move out?

9. Tim Lowell, owner of The Lunch Box Café, has 58 pounds of ground beef on Friday afternoon. If Tim bought another 178 pounds, how much ground beef would he have for the weekend?

10. Chi Yi has these business math test scores for this term: 88, 89, 95, 94, 91, 87. What is Chi's total number of points?

11. Ben Stein sells books. Last week, he drove these many miles:

Monday . . . . .124 miles
Tuesday . . . . .198 miles
Wednesday . . . . . . .72 miles
Thursday . . . . .157 miles
Friday . . . . .205 miles

How many total miles did Ben drive during the week?

12. Itasca Company, which has a chain of grocery stores, showed these sales for the last three months:

western region $345,800
northeastern region $542,875
midwestern region $234,600
southeastern region $423,545

What total sales did the company have for these months?

13. Complete this sales report for Henderson Products:

HENDERSON PRODUCTS
Weekly Sales Report
May 2–6, 20XX

Salesperson	Monday	Tuesday	Wednesday	Thursday	Friday	Total
Booker	$ 876	$1,200	$1,456	$ 978	$1,875	$_____
Vouger	1,250	850	1,245	1,365	1,345	_____
Simmons	1,436	1,243	987	963	1,422	_____
O'Malley	997	1,233	1,577	1,678	1,003	_____
Lang	1,375	1,876	1,986	998	1,245	_____
Sterling	1,855	1,569	1,998	1,755	2,050	_____
Total	+	+	+	+	= _____	

14. Gateway Products Company sold the following numbers of items during the first 6 months of 20XX. Find the total number of items that were sold and fill out the sales report:

GATEWAY PRODUCTS COMPANY
Sales Report
For Period Ending June 30, 20XX

Item	January	February	March	April	May	June	Total
A-9	550	986	376	588	612	618	_____
B-12	890	854	789	987	980	897	_____
B-11	998	887	912	876	670	698	_____
C-1	786	764	876	981	576	886	_____
D-14	312	413	234	543	254	256	_____
D-18	121	213	122	139	298	276	_____
Total	+	+	+	+	+	= _____	

In each of the following application problems, look for combinations that total 10.

15. Southside Grocery made these sales of apples during the past week:

Monday . . . . .48 lb
Tuesday . . . . .62 lb
Wednesday . . . . .75 lb
Thursday . . . . .35 lb
Friday . . . . .88 lb
Saturday . . . . .92 lb

What was the total number of apples sold during the week?

16. Finish this sales report for Kite Company:

KITE COMPANY
Sales Report

Salesperson	Monday	Tuesday	Wednesday	Thursday	Friday
S. Myers	$ 882	$ 978	$ 565	$ 763	$ 762
T. Raja	228	137	545	647	748
G. Goode	327	456	376	765	376
H. Danzio	783	654	934	345	678
Total	$ ___	$ ___	$ ___	$ ___	$ ___

17. Finish filling in this production schedule:

Product		Units Produced			
	Monday	Tuesday	Wednesday	Thursday	Friday
A	167	128	736	891	387
X	943	182	287	298	776
H	487	298	198	187	761
M	428	287	912	923	876
O	222	823	198	187	234
Totals					

Rounding Off Whole Numbers and Estimating Sums

After studying Unit 1.3, you should be able to:

1. Round whole numbers to a specified place.
2. Estimate a sum using rounded numbers.

Astronomers tell us that the sun is 93,000,000 miles from earth. Does this mean that the distance has been exactly measured and the sun is precisely 93,000,000 miles away? Or, does it mean that by using mathematical formulas, astronomers have calculated that the sun is around 93,000,000 miles away and this figure is rounded to even millions to make it easier to work with and to remember?

Actually, the sun is *around 93,000,000* miles from earth, not *exactly* 93,000,000 miles. There are many times when numbers are rounded to make them easier to use. For example, the population of New York City is often said to be 8,000,000 people. At a given time, however, the actual population could be a little over 8,000,000 or a little under 8,000,000. What if it were, say, 8,124,512? This number would be hard to remember. So, it is simply rounded to an even 8,000,000.

To *round off* a number means to use an approximate value instead of the exact value. To round off a number, follow these steps:

Step 1: Locate the place you want to round the number to.

Step 2: Look at the digit to the immediate right of the digit you want to round:

If the digit to the right of the place being rounded is 5 (or more), add 1 to the place being rounded.

If the digit to the right of the place being rounded is less than 5, do not change the digit in the place being rounded.

Step 3: Drop all digits after the place being rounded and replace them with zeros.

Example 1

Round 485 to the nearest ten.

Solution

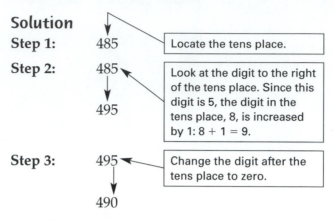

Step 1:　485　— Locate the tens place.

Step 2:　485 → 495　— Look at the digit to the right of the tens place. Since this digit is 5, the digit in the tens place, 8, is increased by 1: 8 + 1 = 9.

Step 3:　495 → 490　— Change the digit after the tens place to zero.

Rounded answer = 490

Example 2

Round 77,242 to the nearest thousand.

Solution

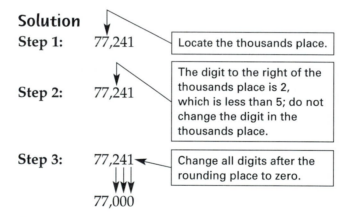

Step 1:　77,241　— Locate the thousands place.

Step 2:　77,241　— The digit to the right of the thousands place is 2, which is less than 5; do not change the digit in the thousands place.

Step 3:　77,241 → 77,000　— Change all digits after the rounding place to zero.

Rounded answer = 77,000

Let's look at a few more examples.

Examples

788 rounded to the nearest ten　　　　　　　　= 790
7,455 rounded to the nearest hundred　　　　　= 7,500
12,898 rounded to the nearest thousand　　　　= 13,000
255,790 rounded to the nearest ten-thousand　= 260,000
397,800 rounded to the nearest hundred-thousand　= 400,000
1,897,476 rounded to the nearest million　　　= 2,000,000

REMEMBER

When rounding, you only look at two digits: the place being rounded and the place to its immediate right; do not look at any other digit.

ESTIMATING SUMS

Have you ever run into your local supermarket to buy a couple of items, ended up buying several, and then held your breath—and prayed that you have enough money—as the cashier rang up the purchases? Most of us have had such an experience. One thing you can do in such a situation is to estimate a sum mentally. To do this, you round each addend to the place of its first digit and then add the rounded amounts. Let's take an example.

Example 3

Estimate the sum of 485 + 211 + 96 + 589.

Solution

	485 rounds to	500
	211 rounds to	200
	96 rounds to	100
	+ 589 rounds to	+ 600
Actual sum ⟶ 1,381		1,400 ◀— Estimated sum

unit 1.3

Name _____ Date _____

Exercises

Round these numbers to the nearest ten:

1. 75

2. 874

3. 999

4. 899

5. 3,456

6. 9,593

7. 12,356

8. 15,837

9. 25,925

10. 33,121

Round these numbers to the nearest hundred:

11. 678

12. 234

13. 1,565

14. 1,843

15. 2,877

16. 4,610

17. 12,875

18. 17,008

19. 21,081

20. 34,571

ANSWERS

1. _____

2. _____

3. _____

4. _____

5. _____

6. _____

7. _____

8. _____

9. _____

10. _____

11. _____

12. _____

13. _____

14. _____

15. _____

16. _____

17. _____

18. _____

19. _____

20. _____

Round these numbers to the nearest thousand:

21. 1,898

22. 2,458

23. 7,916

24. 5,718

25. 9,469

26. 21,967

27. 45,854

28. 67,858

29. 91,546

30. 77,999

Round these numbers to the nearest ten-thousand:

31. 34,598

32. 56,987

33. 47,698

34. 59,308

35. 59,287

36. 18,387

37. 48,698

38. 187,487

39. 287,376

40. 155,985

Round these numbers to the nearest hundred-thousand:

41. 128,765

42. 768,876

43. 345,986

44. 234,765

45. 998,475

46. 705,982

47. 3,409,876

48. 4,986,487

49. 7,856,387

50. 8,265,973

21. _____
22. _____
23. _____
24. _____
25. _____
26. _____
27. _____
28. _____
29. _____
30. _____
31. _____
32. _____
33. _____
34. _____
35. _____
36. _____
37. _____
38. _____
39. _____
40. _____
41. _____
42. _____
43. _____
44. _____
45. _____
46. _____
47. _____
48. _____
49. _____
50. _____

Round these numbers to the nearest million:

51. 3,598,587   **56.** 12,398,459

52. 4,598,219   **57.** 24,998,765

53. 8,498,986   **58.** 56,985,876

54. 3,999,876   **59.** 55,267,985

55. 7,987,376   **60.** 79,999,000

51. _____

52. _____

53. _____

54. _____

55. _____

56. _____

57. _____

58. _____

59. _____

60. _____

Estimate the answer to each of the following problems by rounding each number to its first digit and adding the estimates:

61. Actual	Estimate	**62.** Actual	Estimate	**63.** Actual	Estimate
45		79		124	
71		7		348	
59		60		285	
66		12		810	
54		89		681	
38		56		312	

64. Actual	Estimate	**65.** Actual	Estimate	**66.** Actual	Estimate
459		2,455		7,388	
598		3,599		3,575	
498		4,527		6,810	
482		9,174		5,789	
192		4,585		6,972	
385		6,593		5,025	

Applications

Name _____   Date _____

1. Algood Company had these sales figures for the year:

Territory	Sales
North	$2,988,964
South	3,499,437
East	3,288,388
West	4,988,876
East Central	2,388,984

Round each figure to the nearest thousand dollars.

2. Below are this month's sales for Lakeside Grocery. Round each amount to the nearest ten-thousand dollars:

Grocery: $78,988
Meat: $54,987
Produce: $32,876
Dairy: $18,769
Hardware: $12,344

3. Round the cost of these clothing items to the nearest ten dollars:

Leather coats: $99
Suede coats: $88
Belts: $12
Dress shoes: $65
Top coats: $95

4. Zen Corporation had these profit figures for the last 5 years. Round each figure to the nearest million dollars:

19X1: $2,488,987
19X2: $6,487,184
19X3: $7,765,548
19X4: $7,687,175
19X5: $7,123,465

ANSWERS

1. _____
_____
_____
_____
_____

2. _____
_____
_____
_____
_____

3. _____
_____
_____
_____
_____

4. _____
_____
_____
_____
_____

5. Round 12,467,886 to the nearest:

   a. ten

   b. hundred

   c. thousand

   d. ten-thousand

   e. hundred-thousand

   f. million

5. a. _____

  b. _____

  c. _____

  d. _____

  e. _____

  f. _____

Work Space

Subtracting Whole Numbers and Estimating Differences

After completing Unit 1.4, you will be able to:

1. Find the difference between two numbers.
2. Check the accuracy of a subtraction problem.
3. Estimate the difference between two numbers.

Subtraction is an arithmetic operation in which you find the difference between two numbers. The number from which another number is subtracted (or *taken away*) is called the **minuend**, and the number that is subtracted from the minuend is the **subtrahend.** The answer is called the **difference** or the **remainder.**

You show subtraction by writing the *minus* symbol ("−") in front of the subtrahend. Let's look at a simple problem to review these terms:

$$
\begin{array}{r}
8 \\
-3 \\
\hline
5
\end{array}
\quad
\begin{array}{l}
\longleftarrow \text{minuend} \\
\longleftarrow \text{subtrahend} \\
\longleftarrow \text{difference}
\end{array}
$$

When setting up a subtraction problem (such as with addition), be sure to line up the digits correctly: ones digits over ones digits, tens digits over tens digits, hundreds digits over hundreds digits, and so on. Let's look at two other examples.

Example 1

Properly line up and subtract: 475 − 261.

Solution

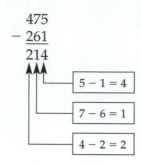

$$
\begin{array}{r}
475 \\
-\ 261 \\
\hline
214
\end{array}
$$

5 − 1 = 4

7 − 6 = 1

4 − 2 = 2

Example 2

Subtract: 8,756 − 445.

Solution

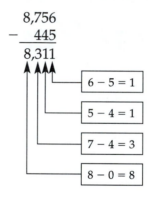

$$
\begin{array}{r}
8,756 \\
-\ \ \ 445 \\
\hline
8,311
\end{array}
$$

6 − 5 = 1

5 − 4 = 1

7 − 4 = 3

8 − 0 = 8

Note that there is no digit in the thousands place of the subtrahend. So, 8 minus zero equals 8.

≈ CHECKING FOR ACCURACY IN A SUBTRACTION PROBLEM

You can check for accuracy in a subtraction problem by adding the difference to the subtrahend; the answer will equal the minuend if your work is correct. For example, we can check the accuracy of the answer in Example 2 as follows:

$$
\begin{array}{rl}
8,311 & \text{difference} \\
+\ \ \ 445 & \text{subtrahend} \\
\hline
8,756 & \text{minuend}
\end{array}
$$

Since the difference—8,311—plus the subtrahend—445—equals the minuend—8,756—our arithmetic is correct.

Self-Check

Subtract the following and check for accuracy:

(a) 558	(b) 57	(c) 569	(d) 4,587
− 12	− 6	− 323	− 2,371

Solution

(a) 546   (b) 51   (c) 246   (d) 2,216

≈≈≈ SUBTRACTING WHEN BORROWING IS NEEDED

Subtraction problems such as those in Examples 1 and 2 usually give us little trouble because each digit in the minuend (the top number) is larger than the corresponding digit in the subtrahend. But, when a digit in the subtrahend is larger than the corresponding digit in the minuend, you must borrow from the digits on the left. For example,

$$\begin{array}{r} 972 \\ -\ 458 \\ \hline ? \end{array}$$ You cannot subtract 8 from 2; so, you must borrow from the "7" in the column to the left.

We can subtract by using the technique of borrowing as follows:

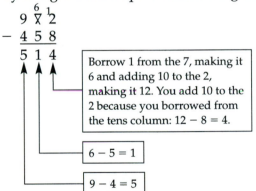

$$\begin{array}{r} 9\ \overset{6}{\cancel{7}}\ \overset{1}{2} \\ -\ 4\ 5\ 8 \\ \hline 5\ 1\ 4 \end{array}$$

Borrow 1 from the 7, making it 6 and adding 10 to the 2, making it 12. You add 10 to the 2 because you borrowed from the tens column: 12 − 8 = 4.

6 − 5 = 1

9 − 4 = 5

Let's look at two additional examples.

Example 1

Subtract and check: $634 - 395$.

Solution

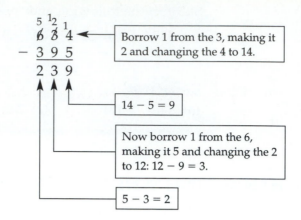

Check

$$
\begin{array}{r}
395 \quad \text{subtrahend} \\
+\ \underline{239} \quad \text{difference} \\
634 \quad \text{minuend}
\end{array}
$$

Example 2

Subtract and check: $4{,}528 - 576$.

Solution

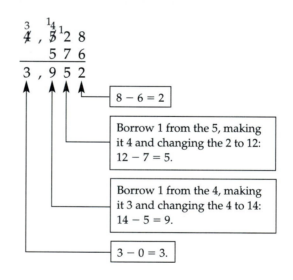

Check

$$
\begin{array}{r}
3{,}952 \quad \text{subtrahend} \\
+\ \underline{576} \quad \text{difference} \\
4{,}528 \quad \text{minuend}
\end{array}
$$

≋ BORROWING FROM ZERO

In many subtraction problems, the digit you need to borrow from will be a 0. When this happens, you must move to the left until you find a digit that is not 0 and borrow from that digit.

Example 1

Subtract: $602 - 428$.

Solution

First, set up the problem correctly:

$$
\begin{array}{r}
602 \\
- \; 428 \\
\end{array}
$$

Now you need to borrow so that the 2 becomes 12. But, you cannot borrow from the 0. So, move over to the 6; borrow 1 from the 60: $60 - 1 = 59$.

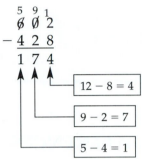

Check

$$
\begin{array}{r}
428 \\
+ \; 174 \\
\hline
602 \\
\end{array}
$$

Example 2

Subtract: $8,007 - 548$.

Solution

Set up the problem correctly:

$$
\begin{array}{r}
8,007 \\
- \quad 548 \\
\end{array}
$$

You must borrow so that the 7 becomes 17. Move to the left to the 8 and borrow 1 from the 800: $800 - 1 = 799$:

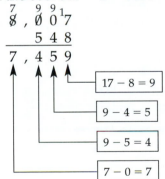

Check

$$7,459$$
$$+ \quad 548$$
$$8,007$$

ESTIMATING DIFFERENCES

We can estimate the difference between two numbers by rounding each number to its largest place and then finding the difference between the two rounded numbers.

Example

Estimate the difference: 10,200 − 3,968. Then find the actual difference.

Solution

Estimate	Actual
10,000	10,200
− 4,000	− 3,968
6,000	6,232

unit 1.4

Exercises

Subtract the following and check for accuracy:

1. 68	**2.** 26	**3.** 65	**4.** 88	**5.** 95	**6.** 25	**7.** 569	**8.** 947
7	21	42	23	75	12	325	205

9. 899	**10.** 805	**11.** 12,345	**12.** 18,549	**13.** 76,008	**14.** 896,995
722	677	2,235	7,455	12,869	875,676

Subtract each of these problems as shown:

15. 18 − 5 = _____   **16.** 21 − 6 = _____   **17.** 56 − 9 = _____   **18.** 87 − 12 = _____

19. 95 − 15 = _____   **20.** 64 − 36 = _____   **21.** 88 − 31 = _____   **22.** 92 − 65 = _____

23. 7,543 − 210 = _____   **24.** 12,455 − 1,126 = _____   **25.** 18,315 − 13,445 = _____

Find the actual and estimated differences between each of the following:

26. **Actual**   **Estimate**   27. **Actual**   **Estimate**
     39,768   18,617
   − 21,332   − 12,910

28. **Actual**   **Estimate**   29. **Actual**   **Estimate**
     24,200   77,812
   − 8,115   − 39,218

30. **Actual**   **Estimate**   31. **Actual**   **Estimate**
     31,165   99,300
   − 7,950   − 69,712

Name _____ Date _____

Applications

1. Complete the following sales report:

MIKE'S SHOES Sales Report October 31, 20XX			
Item	**Beginning Inventory**	**Ending Inventory**	**Units Sold**
Brown casuals	798	312	_____
Nike runners	850	125	_____
Cross trainers	560	78	_____
Dress leather	250	110	_____
No. 2 hikers	128	12	_____
Leather casuals	249	168	_____

2. Complete the following receivables report:

RIVER CITY DEPARTMENT STORE Receivables Report					
Customer	**Purchases**	**−**	**Returns**	**=**	**Balance of Account**
Ben Jacobs	$ 195		$ 45		$_____
Sue Romans	318		125		_____
James Stevens	87		–0–		_____
Bill Willis	450		85		_____
Lisa Young	79		8		_____

3. Bluff City Supply Company offers a discount to all customers who pay their account within 10 days. Complete the following report, which shows all customers who paid early during June:

Account Number	Amount of Purchase	Discount	Amount Received
10018	$12,300	$250	$_____
10036	3,558	65	_____
10045	8,572	160	_____
10052	10,500	210	_____
10065	4,542	92	_____

4. Complete the following inventory sheet and check your answers:

Department	Beginning Inventory	−	Ending Inventory	=	Amount Sold
Men's clothes	$38,560		$22,410		$_____
Women's clothes	55,300		24,678		_____
Shoes	24,645		18,592		_____
Sporting goods	82,598		32,456		_____
Children's	35,608		21,511		_____

5. These tables show the selling price of certain items, and the cost of the same items, at McLean's Department Store. Find the amount of profit (selling price minus cost) on each item and the total profit:

(a)

MCLEAN'S DEPARTMENT STORE Men's Clothing			
Item	Selling Price	Cost	Profit
Leather coats	$195	$125	$_____
Sport coats	99	59	_____
Wool ties	19	10	_____
Silk ties	29	15	_____
Dress shoes	35	21	_____
Wool pants	75	44	_____
Cotton pants	38	22	_____
Belts	18	12	_____
Totals	$_____ −	$_____ =	$_____

(b)

MCLEAN'S DEPARTMENT STORE Women's Clothing			
Item	Selling Price	Cost	Profit
Blouses	$ 37	$ 22	$_____
Coats	199	107	_____
Dresses	49	25	_____
Pants	29	18	_____
Hats	69	52	_____
Leather purses	99	74	_____
Skirts	38	24	_____
Shoes	49	28	_____
Totals	$_____ −	$_____ =	$_____

section

II

Whole Numbers: Multiplication and Division

Multiplying Whole Numbers

After completing Unit 2.1, you will be able to:

1. Multiply whole numbers.
2. Multiply whole numbers using a table.

Multiplication is a shortcut to adding numbers. When you multiply numbers, you are actually adding groups of numbers. For example, multiplying 3 times 4, whose product is 12, is the same as adding 4 together three times: $4 + 4 + 4 = 12$.

Let's look at how a multiplication problem is set up and review its terms:

$$
\begin{array}{r}
23 \\
\times\ \underline{3} \\
69
\end{array}
\begin{array}{l}
\leftarrow\!\!\!-\!\!\!-\!\!\!-\ \text{multiplicand} \\
\leftarrow\!\!\!-\!\!\!-\!\!\!-\ \text{multiplier} \\
\leftarrow\!\!\!-\!\!\!-\!\!\!-\ \text{product}
\end{array}
$$

Note that, as in addition and subtraction, multiplication problems are set up by correctly lining up digits: ones digits over ones digits, hundreds digits over hundreds digits, and so on. Also, note that the number being multiplied is called the **multiplicand** and the number by which you multiply the multiplicand is called the **multiplier.** The answer in a multiplication problem is called the **product.**

Let's look at another example.

Example
Multiply 345×4.

Solution
Step 1: Set up the problem correctly:

$$
\begin{array}{r}
345 \\
\times\ \underline{4}
\end{array}
$$

Step 2: Start multiplying from right to left: $4 \times 5 = 20$. Write 0 and carry 2 to the top of the next column:

$$
\begin{array}{r}
2 \quad \longleftarrow \text{carry} \\
345 \\
\times \quad 4 \\
\hline
0
\end{array}
$$

Step 3: Continue multiplying: $4 \times 4 = 16$, plus the 2 you carried equals 18. Write 8 and carry 1 to the top of the next column:

$$
\begin{array}{r}
12 \quad \longleftarrow \text{carry} \\
345 \\
\times \quad 4 \\
\hline
80
\end{array}
$$

Step 4: Finish doing your multiplication: $4 \times 3 = 12$, plus the 1 you carried equals 13. Write 13:

$$
\begin{array}{r}
12 \\
345 \\
\times \quad 4 \\
\hline
1,380
\end{array}
$$

 ## USING A MULTIPLICATION TABLE

The best way to improve your speed and accuracy when multiplying is to be sure that you know your multiplication tables. The following table can be used to practice your basic multiplication skills:

Basic Multiplication Facts

×	0	1	2	3	4	5	6	7	8	9
0	0	0	0	0	0	0	0	0	0	0
1		1	2	3	4	5	6	7	8	9
2			4	6	8	10	12	14	16	18
3				9	12	15	18	21	24	27
4					16	20	24	28	32	36
5						25	30	35	40	45
6							36	42	48	54
7								49	56	63
8									64	72
9										81

To illustrate how to use the table, let's say that you want to multiply 7×8. First, find the 7 *under* the column marked "X" at the extreme left. Then find the 8 in the very top row to the right of the "X." Read across from the 7 and down from the 8 until you meet; you will see the number 56. So, $7 \times 8 = 56$.

Notice that it is not necessary to show all products in the table. This is because 7×8 is the same as 8×7; the order of multiplication does not matter. Also, notice that multiplying by 1 does not change the other number and that multiplying by zero always gives a product of zero.

MULTIPLYING NUMBERS WITH TWO OR MORE DIGITS IN THE MULTIPLIER

When multiplying numbers, you will often work with multipliers that have two or more digits. When this happens, you multiply the multiplicand by *each* digit in the multiplier. Each separate multiplication gives you a **partial product.** You then add the partial products to get the final product.

Example

Multiply: 645×35.

Solution

Step 1: Set up the problem correctly (remember, ones digits over ones digits, tens digits over tens digits, and so on):

$$\begin{array}{r} 645 \\ \times\ \ 35 \\ \hline \end{array}$$

Step 2: Multiply the multiplicand by the ones digit in the multiplier (5, in this problem):

$$\begin{array}{r} 22 \\ 645 \\ \times\ \ \ 35 \\ \hline 3{,}225 \end{array}$$ ◄—— partial product: $5 \times 645 = 3{,}225$

Step 3: Multiply the multiplicand by the tens digit in the multiplier (3, in this problem). You always put the first digit in a partial product right under the number you multiplied with. So, write the first digit in this partial product right under the 3:

$$\begin{array}{r} 11 \\ 645 \\ \times\ \ \ \ 35 \\ \hline 3225 \\ 1935 \end{array}$$ ◄—— partial product: $3 \times 645 = 1{,}935$

Step 4: Add the partial products to get the final product:

$$\begin{array}{r} 645 \\ \times\ \ \ \ 35 \\ \hline 3225 \\ 1935\ \ \\ \hline 22575 \end{array}$$ ◄—— product

Follow these same steps if you are multiplying larger numbers. Remember that each partial product is always written one place to the left of the partial product before it. Let's look at one more example.

Example
Multiply: $3,455 \times 324$.

Solution
Step 1: Set up the problem correctly:

$$
\begin{array}{r}
3,455 \\
\times 324 \\
\hline
\end{array}
$$

Step 2: Multiply the multiplicand by the ones digit in the multiplier (4, in this problem):

$$
\begin{array}{r}
1\,22 \\
3,455 \\
\times 324 \\
\hline
13,820
\end{array}
\qquad \boxed{4 \times 3,455 = 13,820}
$$

Step 3: Multiply the multiplicand by the tens digit in the multiplier (2, in this problem):

$$
\begin{array}{r}
1\,11 \\
3,455 \\
\times 324 \\
\hline
13820 \\
6910
\end{array}
\qquad \boxed{2 \times 3,455 = 6,910}
$$

Step 4: Multiply the multiplicand by the hundreds digit in the multiplier (3, in this problem):

$$
\begin{array}{r}
1\,11 \\
3,455 \\
\times 324 \\
\hline
13820 \\
6910 \\
10365
\end{array}
\qquad \boxed{3 \times 3,455 = 10,365}
$$

Step 5: Add the partial products to get the product:

$$
\begin{array}{r}
3,455 \\
\times 324 \\
\hline
13820 \\
6910 \\
10365 \\
\hline
1119420 = 1,119,420
\end{array}
$$

MULTIPLYING WITH ZEROS IN THE MULTIPLIER

You should be careful when your multiplier has a zero (or zeros) in it because, as we discussed earlier, any number multiplied by zero equals zero. Let's look at an example of multiplication when the multiplier has a zero.

Example

Multiply 265×205.

Solution

$$
\begin{array}{r}
265 \\
\times \quad 205 \\
\hline
1,325 \quad \longleftarrow \quad 5 \times 265 = 1,325 \\
000 \quad \longleftarrow \quad 0 \times 265 = 000 \\
530 \quad \longleftarrow \quad 2 \times 265 = 530 \\
\hline
54,325
\end{array}
$$

While this method works well, it often causes errors because we can easily forget that zero times any number is zero. We can make the multiplication easier by not multiplying by the zero. Instead, just bring down the 0 into the product as a **placeholder.** Let's look at a couple of examples to see how this works.

Example 1

Multiply: 265×205.

Solution

Step 1: Set up the problem correctly and multiply the multiplicand by the ones digits in the multiplier:

$$
\begin{array}{r}
32 \\
265 \\
205 \\
\hline
1,325 \quad \longleftarrow \quad 5 \times 265 = 1,325
\end{array}
$$

Step 2: Bring down the zero into the product as a placeholder:

$$
\begin{array}{r}
265 \\
205 \\
\hline
1,325 \\
0
\end{array}
$$

Step 3: Multiply the multiplicand by the hundreds digit in the multiplier:

$$
\begin{array}{r}
11 \\
265 \\
\underline{205} \\
1325 \\
\underline{5300} \quad
\end{array}
$$
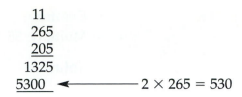
$\longleftarrow 2 \times 265 = 530$

Step 4: Add the partial products:

$$
\begin{array}{r}
265 \\
\underline{205} \\
1325 \\
\underline{5300} \\
54{,}325
\end{array}
$$

Example 2

Multiply: $5{,}225 \times 2{,}004$.

Solution

Step 1: Multiply the multiplicand by the ones digit in the multiplier:

$$
\begin{array}{r}
12 \\
5225 \\
\underline{2004} \\
20{,}900
\end{array}
$$
$\longleftarrow 4 \times 5{,}225 = 20{,}900$

Step 2: Bring down the zeros into the product as placeholders:

$$
\begin{array}{r}
5{,}225 \\
\underline{2{,}004} \\
20900 \\
00
\end{array}
$$

Step 3: Multiply the multiplicand by the thousands digit in the
   multiplier:

$$
\begin{array}{r}
1 \\
5{,}225 \\
\underline{2{,}004} \\
20900 \\
1045000
\end{array}
$$
$\longleftarrow 2 \times 5{,}225 = 10{,}450$

Step 4: Add the partial products:

$$
\begin{array}{r}
5{,}225 \\
\underline{2{,}004} \\
20900 \\
\underline{1045000} \\
10{,}470{,}900
\end{array}
$$

MULTIPLYING BY A POWER OF TEN

When your multiplier is 10 or a power of 10 (for example, 100, 1,000, 10,000, etc.)
and your multiplicand is a whole number, you can shortcut multiply by counting the zeros in the multiplier and adding on the same number of zeros to the
multiplicand.

Example
Multiply: 55 × 10.

Solution

$$55 \times 10 = 550$$

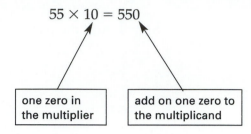

| one zero in the multiplier | add on one zero to the multiplicand |

Let's look at a few more examples to see how well this works with larger numbers.

Examples
Multiply the following:
a. 657 × 1,000
b. 1,837 × 10,000
c. 12,456 × 100
d. 55,686 × 100,000

Solution
a. 657 × 1,000 = 657,000 (add on three zeros)
b. 1,837 × 10,000 = 18,370,000 (add on four zeros)
c. 12,456 × 100 = 1,245,600 (add on two zeros)
d. 55,686 × 100,000 = 5,568,600,000 (add on five zeros)

 ## MULTIPLYING WHEN THE MULTIPLIER ENDS WITH ZEROS

When the multiplier ends with a zero (or zeros), you can avoid multiplying by the zeros if you follow the steps in the next example.

Example
Multiply: 515 × 200

Solution
Step 1: Set the zeros to the right of the multiplicand:

$$\begin{array}{r} 515 \\ \underline{200} \end{array}$$

Step 2: Bring zeros into the product:

$$\begin{array}{r} 515 \\ \underline{200} \\ 00 \end{array}$$

Step 3: Multiply by the nonzero whole numbers:

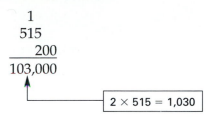

$$
\begin{array}{r}
1 \\
515 \\
\underline{200} \\
103{,}000
\end{array}
$$

2 × 515 = 1,030

Exercises

Name _____ Date_____

Multiply each of these problems:

1. 45 2	**2.** 65 5	**3.** 69 8	**4.** 75 3	**5.** 26 6	**6.** 98 9	**7.** 54 2

8. 65 4	**9.** 76 7	**10.** 87 8	**11.** 58 7	**12.** 48 5	**13.** 39 9	**14.** 49 4

15. 235 5	**16.** 455 9	**17.** 983 6	**18.** 285 8	**19.** 584 7	**20.** 298 9

21. 4,594 5	**22.** 5,938 7	**23.** 6,398 8	**24.** 67,398 5	**25.** 34,877 9

Using the table shown in this unit, multiply these numbers:

26. $8 \times 9 =$ _____   **27.** $5 \times 8 =$ _____   **28.** $7 \times 7 =$ _____   **29.** $6 \times 9 =$ _____

30. $9 \times 9 =$ _____   **31.** $9 \times 0 =$ _____   **32.** $4 \times 8 =$ _____   **33.** $5 \times 7 =$ _____

34. $0 \times 9 =$ _____

Multiply each of these problems:

35. 46
25

36. 38
17

37. 59
49

38. 39
26

39. 98
84

40. 85
28

41. 87
27

42. 97
82

43. 387
80

44. 598
86

45. 307
45

46. 497
68

47. 298
38

48. 398
85

49. 985
33

50. 4,599
241

51. 5,388
208

52. 4,288
600

53. 9,451
298

54. 12,300
500

55. 54,388
200

56. 23,487
492

57. 36,871
593

58. 125,601
4,555

59. 218,452
10,005

60. 554,693
15,638

61. $83 \times 6,000 =$

62. $937 \times 500 =$

63. $1,564 \times 200 =$

64. $25,692 \times 900 =$

65. $8,436 \times 700 =$

Mentally multiply the following:

66. $561 \times 10 =$

67. $345 \times 100 =$

68. $481 \times 1,000 =$

69. $2,488 \times 100 =$

70. $459 \times 10,000 =$

71. $3,498 \times 100,000 =$

72. $24,588 \times 1,000 =$

73. $125,675 \times 100 =$

74. $300 \times 50 =$

ANSWERS

61. _____

62. _____

63. _____

64. _____

65. _____

66. _____

67. _____

68. _____

69. _____

70. _____

71. _____

72. _____

73. _____

74. _____

unit 2.1

Applications

1. Bill Tai bought 8 pounds of peanuts at $2 a pound. How much did he spend?

2. A local baseball stadium has 7,552 seats. At $3 a ticket, how much money would be received on the night of a sellout?

3. If a crate of oranges holds 756 oranges, how many oranges will 8 crates hold?

4. If John Sansone puts $9 per week in his Christmas Club account for a whole year, how much will he have in the account by the end of the year?

5. If Bob Shiffman makes $12 an hour, what would he make in a week in which he worked 38 hours?

6. If a certain brand of candy costs $16 a pound. How much would 15 pounds cost?

7. If Ann's salary is $1,575 per month, what is her salary for a year?

8. A local sporting goods store bought 25 pairs of jogging shoes. If the price of each pair was $30, how much did the store pay to buy all of them?

9. A local boy's club is having a cookout. They need 15 pounds of hamburger meat at $2 a pound, 25 packs of buns at $1 each, 8 packs of potato chips at $1 each, and 4 large bottles of cola at $2 each. What will be the cost of the cookout?

10. There are 500 sheets in a ream of typing paper. How many sheets are in 50 reams?

11. If envelopes are packaged 1,000 to a box, how many envelopes are in 50 boxes?

12. During a clearance sale, Duke Electronics sold 2,455 video recorders for $1,005 each. What was the total amount received for all sales?

ANSWERS

1. _____
2. _____
3. _____
4. _____
5. _____
6. _____
7. _____
8. _____
9. _____
10. _____
11. _____
12. _____

13. Complete this daily earnings sheet for Madison Company.

MADISON COMPANY
Daily Earnings Sheet

Employee	Hourly Wage	Hours Worked	Total Earnings
J. Myers	$ 8	8	$____
H. Booke	9	7	____
D. Swartz	12	8	____
K. Burke	10	6	____
B. Lasser	15	8	____
C. Bauer	11	7	____
T. Culman	14	8	____
B. Mayo	15	8	____
		Total	$____

14. Fill out this purchase order. *Note:* The symbol "@" means "at."

PURCHASE ORDER

To: **Allied Office Supply Co.**   No. __472__

Manchester, N.H.   Date: __8/18/20XX__

Quantity	Description	Amount
200 reams	20 lb bond @ $18 each	
18 dozen	Yellow writing pads @ $12 dz	
12 dozen	#3 ballpoint pens @ $3 dz	
1	Desktop calculator @ $125 each	
15	#125A computer ribbons @ $12 each	
18 boxes	Correction fluid @ $5 box	
	Total	

15. Complete this invoice:

VICTORY AUTO PARTS COMPANY
Invoice

To: **Bill Stoops**   Date: **11-6-20XX**

Terms: **Cash**

Quantity	Description	Unit Price	Amount
50	#8 spark plugs	$ 2	$_____
75	Rearview mirrors	12	_____
90	Sideview mirrors	18	_____
15	#120A blackwall tires	35	_____
20	#128B whitewall tires	42	_____
25	#45D floor mats	21	_____
80	#2 deluxe seat cushions	9	_____
		Total $	

16. Fill out this sales report for the Rosenberg Company:

ROSENBERG COMPANY
Sales Report

Product	Number of Units Sold	Unit Selling Price	Total Sale
A	200	$ 18	$_____
B	300	12	_____
C	500	8	_____
D	800	5	_____
E	600	15	_____
F	350	10	_____
G	980	100	_____
H	532	200	_____
I	1,255	100	_____
J	1,855	120	_____
Totals	_____		$_____

unit 2.2

Checking Multiplication and Estimating Answers

After completing Unit 2.2, you will be able to:

1. Check for accuracy in multiplication problems by reversing the order of multiplication.
2. Estimate products using rounded numbers.

As we have stressed, the accuracy in any arithmetic operation should always be checked. The best way to check the accuracy of a multiplication problem is simply to reverse the order of the multiplier and multiplicand, and multiply again. As we learned in Unit 2.1, the order of multiplication does not matter. An example will easily illustrate this.

Example
Multiply 325 × 128 and check for accuracy by reverse multiplication.

Solution	Accuracy Check

$$
\begin{array}{r}
325 \\
\times \quad 128 \\
\hline
2600 \\
650 \\
325 \\
\hline
41600
\end{array}
\qquad
\begin{array}{r}
128 \\
\times \quad 325 \\
\hline
640 \\
256 \\
384 \\
\hline
41600
\end{array}
$$

41600 ⟷ 41600

Self-Check

Multiply the following and check your answer: 375 × 218.

Solution

81,750

ESTIMATING ANSWERS IN MULTIPLICATION PROBLEMS

Answers to multiplication problems can be estimated by following these steps:

Step 1: Round the multiplier to the value of its first digit (for example, 321 becomes 300).

Step 2: Round the multiplicand to the value of its first digit (for example, 690 becomes 700).

Step 3: Multiply the rounded numbers. The rounded answer you get will be close to the actual answer.

Example

Multiply: 395 by 39.

Solution

Estimate	Actual
400	395
× 40	× 39
16,000	15,405

Self-Check

Estimate the product of 18,873 × 21. Find the actual product.

Solution

Estimated product: 380,000
Actual product: 396,333

Name _____ **Date** _____

Exercises

unit 2.2

Find the product of each of the following problems. Check your answer by reversing the order of the multiplier and multiplicand, and multiplying again.

1.	Check	2.	Check	3.	Check	4.	Check
75 × 12		385 × 98		714 × 87		440 × 315	

5.	Check	6.	Check	7.	Check	8.	Check
3,456 × 618		2,975 × 405		5,618 × 455		4,125 × 345	

9.	Check	10.	Check	11.	Check	12.	Check
615 × 324		419 × 398		533 × 299		8,256 × 217	

Estimate the product of each of the following. Then compute the actual product:

13.	Estimate	Actual	14.	Estimate	Actual	15.	Estimate	Actual
118 × 92			201 × 87			879 × 314		

16.	Estimate	Actual	17.	Estimate	Actual	18.	Estimate	Actual
1008 × 34			980 × 224			785 × 18		

Name _____ Date_____

Applications

1. Complete the following invoice and check your answers:

	STERLING NATURAL FOODS		
	311 Chestnut Street		
	San Francisco, CA 94111		

To: Alli's Health Foods	Invoice No.	408
1002 Cascade Drive	Date	June 12, 20–
San Francisco, CA 94111	Terms	Cash
	Shipment	Truck

Quantity	Item	Unit Price	Extension
125	12 oz. sea salt	$ 2.00	
15	Exercise bikes	45.00	
10	Treadmills	90.00	
10	Multi-Plex abs	22.00	
12	Waist rollers	15.00	
30	Jump ropes	6.00	
50	Cases - vit. E	95.00	
50	Cases - vit. C	57.00	
75	Cases - mult.	99.00	
90	Cases-kids vit.	65.00	
20	Cases-hi pro	10.00	
35	Cases - dry E	37.00	
		Total _____	

2. The following is a completed invoice. Mentally check the extensions by rounding to the value of the first digit in each number.x

JONES OFFICE SUPPLY
211 Peachtree
Atlanta, GA 30304

To	Ben's Bookkeeping & Tax		Invoice No.	10003
	345 Forrest Avenue		Date	4-28-20XX
	Atlanta, GA 30301		Terms	Cash
			Shipment	UPS

Quantity	Item	Unit Price	Extension
10	#8 ribbons	$8.00	$ 80.00
12	#2 staples	9.00 box	108.00
2	Laser cart.	89.00	89.00
15	Reams #10045	15.00	225.00
18	Reams #10046	9.00	162.00
50	#3122 com. pap.	6.00	300.00
25	Paper out	2.00	50.00
			Total $1,014.00

unit 2.3

Dividing Whole Numbers

After completing Unit 2.3, you will be able to:

1. Divide whole numbers where there is no remainder.
2. Divide whole numbers and express a remainder.

If a pie is cut into six equal pieces to serve six people, we say that the pie has been *divided* six ways. In the same way, if a high-school debate team that has 10 members is given $200 for expenses for a road trip, the $200 must be divided by 10. **Division** is the process of finding out how many times one number (the **divisor**) is contained in another number (the **dividend**). The answer in a division problem is called the **quotient.** Division is the opposite of multiplication. Let's look at a division problem and review its terms.

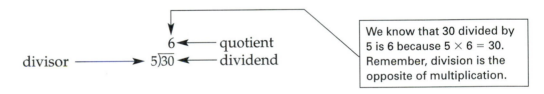

We know that 30 divided by 5 is 6 because 5 × 6 = 30. Remember, division is the opposite of multiplication.

In this example, we use the division box $)\overline{}$ to indicate division. Another common way to indicate division is with the symbol "÷," which means *divided by.* Thus, 30 ÷ 5 means the same as $5\overline{)30}$.

To find the answer to some division problems, you have to use long division. Let's take an example.

Example
Divide 684 by 9.

Solution
Step 1: Set up the problem in a division box:

$$9\overline{)684}$$

Step 2: Look at the divisor (9) and the first digit in the dividend (6). How many 9's are there in 6? None. So, move over to the 8 and ask how many 9's are in 68? There are seven 9's in 68, so you place the 7 directly above the 8:

$$\begin{array}{r} 7 \\ 9\overline{)684} \end{array}$$

$68 \div 9 = 7$

Step 3: Multiply $7 \times 9 = 63$. Write this product under the 68 in the dividend:

$$\begin{array}{r} 7 \\ 9\overline{)684} \end{array}$$

$7 \times 9 = 63 \longrightarrow 63$

Step 4: Subtract 63 from 68 to get a difference of 5 ($68 - 63 = 5$). Bring down the 4 from the dividend and write it next to the 5:

$$\begin{array}{r} 7 \\ 9\overline{)684} \\ 63 \\ \hline 54 \end{array}$$

new dividend

Step 5: Now you have a new dividend, 54. Divide the divisor, 9, into the new dividend. The result is 6 ($54 \div 9 = 6$). Write 6 in the quotient and multiply the 6 by the divisor. Write the result ($6 \times 9 = 54$) under the previous dividend of 54.

$$\begin{array}{r} 76 \\ 9\overline{)684} \\ 63 \\ \hline 54 \\ 54 \\ \hline 0 \end{array}$$

$6 \times 9 = 54$

The last subtraction results in a difference of 0. So, 9 is contained in 684 exactly 76 times. When this happens, we say that the problem "comes out even." If a division problem does not come out even, there will be a **remainder** after you make your last subtraction. We will work with remainders in the next section.

Self-Check

Divide each of the following:

(a) $3\overline{)36}$ (b) $8\overline{)48}$ (c) $4\overline{)844}$ (d) $5\overline{)2,555}$ (e) $45\overline{)360}$

Solution

(a) 12   (b) 6   (c) 211   (d) 511   (e) 8

DIVISION WITH A REMAINDER

In the previous section, we worked with division problems that "came out even." That is, the divisor divided evenly into the dividend without a final remainder. In many division problems, often there will be a remainder after you do the final subtraction. We call this **division with a remainder.**

Example
Divide 658 by 8.

Solution

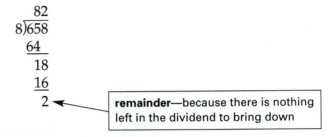

$$\begin{array}{r} 82 \\ 8\overline{)658} \\ 64 \\ \hline 18 \\ 16 \\ \hline 2 \end{array}$$

remainder—because there is nothing left in the dividend to bring down

In this problem, we say that 658 divided by 8 equals 82 with "2 left over." So, the final answer can be stated as 82 r2 (which you read "82 with a remainder of 2"). Another way to show the remainder as part of the answer is to write it as a *fraction*. To show a remainder as a fraction, draw a line under the remainder and write the divisor under the line. Then write the fraction next to the quotient. As an example, let's look again at the problem:

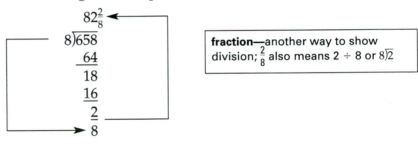

$$\begin{array}{r} 82\frac{2}{8} \\ 8\overline{)658} \\ 64 \\ \hline 18 \\ 16 \\ \hline 2 \\ \hline 8 \end{array}$$

fraction—another way to show division; $\frac{2}{8}$ also means $2 \div 8$ or $8\overline{)2}$

Now let's look at a couple of additional examples.

Example 1
Divide 1,234 by 9 and express the remainder as a fraction.

Solution

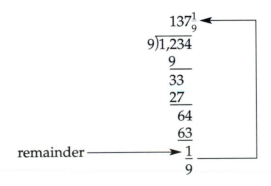

$$\begin{array}{r} 137\frac{1}{9} \\ 9\overline{)1,234} \\ 9 \\ \hline 33 \\ 27 \\ \hline 64 \\ 63 \\ \hline 1 \\ \hline 9 \end{array}$$

remainder

Example 2
Divide 3,562 by 315 and express the remainder as a fraction.

Solution

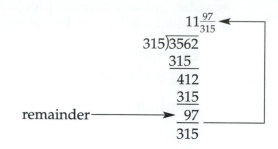

$$11\frac{97}{315}$$
$$315\overline{)3562}$$
$$\underline{315}$$
$$412$$
$$\underline{315}$$
remainder ⟶ $$97$$
$$\overline{315}$$

≋ DIVIDING WHEN THE DIVIDEND ENDS WITH ZEROS

When the dividend ends with a zero (or zeros), just move the zero(s) into the quotient and divide by the nonzero whole numbers, as we see in the next two examples.

Example 1
Divide 4,500 by 9.

Solution

$$45 \div 9 = 5$$

$$9\overline{)4,500} = 500$$
$$500$$

Example 2
Divide 30,000 by 15.

Solution

$$30 \div 15 = 2$$

$$15\overline{)30,000} = 2,000$$
$$2,000$$

DIVIDING WHEN THE DIVISOR AND THE DIVIDEND END WITH ZEROS

When *both* the divisor and the dividend end with zeros, drop the zeros at the end of the divisor, and drop the *same* number of zeros at the end of the dividend; then divide by the nonzero whole numbers.

Example 1

Divide 1,200 by 400.

Solution

$$12 \div 4 = 3$$

$$\begin{array}{r} 3 \\ 4\cancel{00}\overline{)12\cancel{00}} \\ \underline{12} \end{array} \qquad = 3$$

Example 2

Divide 45,000 by 500.

Solution

$$450 \div 5 = 90$$

$$\begin{array}{r} 90 \\ 5\cancel{00}\overline{)450\cancel{00}} \\ \underline{450} \end{array} \qquad = 90$$

Example 3

Divide 36,000 by 60.

Solution

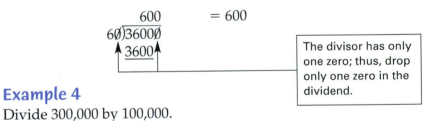

$$\begin{array}{r} 600 \\ 6\cancel{0}\overline{)3600\cancel{0}} \\ \underline{3600} \end{array} \qquad = 600$$

The divisor has only one zero; thus, drop only one zero in the dividend.

Example 4

Divide 300,000 by 100,000.

Solution

$$\begin{array}{r} 3 \\ 1\cancel{00{,}000}\overline{)3\cancel{00{,}000}} \\ \underline{3} \end{array} \qquad = 3$$

Exercises

Divide each of these problems. Show any remainder as a fraction:

1. 4)23 **2.** 8)48 **3.** 4)844 **4.** 5)525 **5.** 7)6,377

6. 8)6,496 **7.** 6)4,284 **8.** 9)3,699 **9.** 7)49,420 **10.** 25)200

11. 68)340 **12.** 25)3,050 **13.** 30)6,330 **14.** 24)8,424 **15.** 214)642

16. 147)735 **17.** 225)1,800 **18.** 301)1,806 **19.** 211)1,055

20. 75)24,375 **21.** 725)6,525 **22.** 636)28,620 **23.** 9)683

ANSWERS

1. _____
2. _____
3. _____
4. _____
5. _____
6. _____
7. _____
8. _____
9. _____
10. _____
11. _____
12. _____
13. _____
14. _____
15. _____
16. _____
17. _____
18. _____
19. _____
20. _____
21. _____
22. _____
23. _____

24. $8\overline{)51}$ **25.** $18\overline{)657}$ **26.** $23\overline{)397}$ **27.** $15\overline{)1,128}$

28. $16\overline{)1,429}$ **29.** $25\overline{)2,378}$ **30.** $21\overline{)1,871}$ **31.** $19\overline{)1,655}$

32. $17\overline{)1,718}$ **33.** $715\overline{)32,322}$ **34.** $313\overline{)29,737}$ **35.** $417\overline{)24,187}$

36. $218\overline{)125,600}$ **37.** $315\overline{)630,945}$ **38.** $12,569 \div 45$ **39.** $91,179 \div 918$

40. $123,125 \div 655$ **41.** $144,516 \div 612$ **42.** $\$16,920 \div 235$ **43.** $\$28,305 \div 629$

44. $\$13,440 \div 384$ **45.** $\$27,170 \div 715$ **46.** $300\overline{)1,200}$ **47.** $800\overline{)64,000}$

48. $200\overline{)14,400}$ **49.** $400\overline{)100,000}$ **50.** $45,000\overline{)135,000}$

51. $5,000\overline{)25,000}$ **52.** $7,800\overline{)1,950,000}$ **53.** $12,000\overline{)1,800,000}$

54. $15,000\overline{)600,000}$ **55.** $200,000\overline{)6,000,000}$

24. _____
25. _____
26. _____
27. _____
28. _____
29. _____
30. _____
31. _____
32. _____
33. _____
34. _____
35. _____
36. _____
37. _____
38. _____
39. _____
40. _____
41. _____
42. _____
43. _____
44. _____
45. _____
46. _____
47. _____
48. _____
49. _____
50. _____
51. _____
52. _____
53. _____
54. _____
55. _____

Name _____ Date _____

Applications

1. Lori Kaye bought a new computer on credit for $1,236. If Lori made 12 equal payments, what was the amount of each payment?

ANSWERS

1. _____

2. _____

3. _____

4. _____

5. _____

2. A local furniture store bought 72 tables for $3,960. What is the cost of each table?

3. If 18 ounces of candy are divided among three children, how much would each child get?

4. A local television station is paying $13,000 for the rights to broadcast a semi-pro baseball game. If there are 20 players on each team (40 players altogether), how much would each player be paid?

5. A motorcycle costing $5,000 can be bought with a $500 down payment and the balance to be paid in 36 monthly payments. What is the amount of each payment?

6. McGram Company pays $27,000 a month for health insurance for its 150 employees. What is the cost for each employee per month?

7. What is the cost of 1 foot of copper tubing if 12 feet cost $24.

8. If a 12-case box of canned fruit weighs 46 pounds, what does each can weigh?

9. It takes 2 yards of fabric to make a certain kind of T-shirt. (a) How many T-shirts can be made with 99 yards of fabric? (b) How much fabric is left over?

10. A six-member jogging team carried 7 quarts of juice with them on a cross-country run. At 32 ounces per quart, how many ounces of juice were there for each runner?

11. Ann Todd makes $325 for a 40-hour week. What does she make an hour?

12. Vicki and Mike Matli paid $7,350 for 600 square yards of carpet. What is the cost per square yard?

13. Ted Davis read that playing racquetball burns 400 calories an hour. How many hours would Ted need to play to burn off a large piece of chocolate cake that contained 1,100 calories?

13. _____

14. _____

15. _____

16. _____

17. _____

14. Five members of the Midtown Boy's Club held a bake sale to raise funds. The sale was a success but there were 16 pieces of pie left over. So, the boys divided up the leftover pies evenly. How much was each boy's share?

15. The Baker family is planning to drive 2,321 miles on their vacation. How many miles will they have to drive each day if they plan to make the trip in 3 days?

16. How long would it take a plane traveling at 200 miles per hour to fly 3,600 miles?

17. There are around 75,000,000 personal computers in use in the United States. Considering a population of about 250,000,000, there is one computer for every _____ people. Please fill in the blank.

18. The job title and yearly salaries for certain jobs in the Goozner Company are shown in the following form. Complete the form:

Job	Yearly Salary	Monthly* Salary	Weekly** Salary
Computer operator	$19,500	$_____	$_____
Computer programmer	39,000	_____	_____
Accountant	23,400	_____	_____
Secretary	11,700	_____	_____
Salesperson	35,100	_____	_____
Sales manager	46,800	_____	_____

*Divide yearly salary by 12.
**Divide yearly salary by 52.

unit 2.4

Dividing When Zeros Are Needed in the Quotient

After completing Unit 2.4, you will be able to:

1. Divide whole numbers when zeros are needed in the quotient.

We have stressed that in our system of place values, a zero contained in a number means no value in that place. For example, in the number 605, the 0 means "no value in the tens place." But even though the 0 means there are no tens, it is still needed as a placeholder or else the total value of the number will be changed. To illustrate this, say that you work in a department store and sold something for $605. However, in writing up the sale, you left out the 0 by mistake and wrote $65. This would be an expensive mistake; instead of charging 6 hundreds, 0 tens, and 5 ones, you charged 6 tens and 5 ones.

In many division problems, a 0 will be needed in the quotient to serve as a placeholder. An example will show when a 0 is necessary.

Example
Divide 6,122 by 15.

Solution
As you have learned, start the division process by dividing 15 into 61:

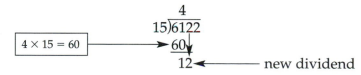

The next step is to divide the new dividend by the divisor. But, note that the new dividend, 12, is smaller than the divisor. So, proceed as follows:

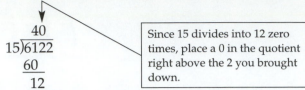

$$\begin{array}{r} 40 \\ 15\overline{)6122} \\ \underline{60} \\ 12 \end{array}$$

Since 15 divides into 12 zero times, place a 0 in the quotient right above the 2 you brought down.

Now form a new dividend by bringing down the next digit:

placeholder

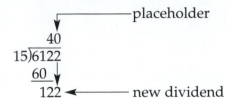

$$\begin{array}{r} 40 \\ 15\overline{)6122} \\ \underline{60} \\ 122 \end{array}$$ ◄——— new dividend

Then divide the new dividend, 122, by the divisor, 15:

$122 \div 15 = 8$

$$\begin{array}{r} 408 \\ 15\overline{)6122} \\ \underline{60} \\ 122 \\ \underline{120} \\ 2 \end{array}$$

$8 \times 15 = 120$

2 ◄——— remainder

Finally, state the remainder as a fraction:

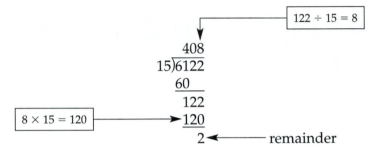

$$\begin{array}{r} 408\frac{2}{15} \\ 15\overline{)6122} \\ \underline{60} \\ 122 \\ \underline{120} \\ \underline{2} \\ 15 \end{array}$$

Now let's look at another example.

Example
Divide 2,472 by 12.

Solution
Start the division process by dividing 12 into 24:

$$\begin{array}{r} 2 \\ 12\overline{)2472} \\ \underline{24} \\ 7 \end{array}$$ ◄——— new dividend

Note that the new dividend, 7, is smaller than the divisor, 12. So, place a 0 in the quotient right above the 7 you brought down:

$$
\begin{array}{r}
20 \\
12\overline{)2472} \\
\underline{24} \\
7
\end{array}
$$

Now bring down the next digit to form a new dividend:

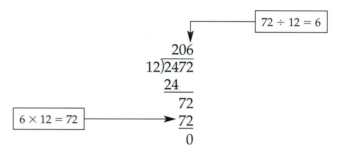

new dividend

Finally, divide the new dividend, 72, by the divisor, 12:

$$72 \div 12 = 6$$

$$
\begin{array}{r}
206 \\
12\overline{)2472} \\
\underline{24} \\
72 \\
72 \\
\underline{72} \\
0
\end{array}
$$

$6 \times 12 = 72$

Since there is no remainder, the problem "comes out even."

Self-Check
Divide the following:

(a) $30\overline{)6{,}120}$ (b) $38\overline{)4{,}104}$ (c) $24\overline{)14{,}424}$ (d) $22\overline{)4{,}558}$

(e) $150\overline{)301{,}050}$

Solution
(a) 204 (b) 108 (c) 601 (d) $207\frac{4}{22}$
(e) 2,007

Exercises

Divide each of these problems. Where needed, put zeros in the quotient:

1. 15)3,060

2. 24)2,549

3. 17)5,219

4. 14)4,270

5. 18)9,090

6. 19)2,052

7. 12)7,212

8. 27)2,838

9. 35)3,815

10. 27)8,262

11. 13)5,239

12. 14)7,014

13. 21)4,368

14. 17)5,440

15. 24)9,672

16. 11)2,279

ANSWERS

1. _____
2. _____
3. _____
4. _____
5. _____
6. _____
7. _____
8. _____
9. _____
10. _____
11. _____
12. _____
13. _____
14. _____
15. _____
16. _____

17. 35)14,315

18. 42)21,336

19. 33)20,031

17. _____

18. _____

19. _____

20. _____

21. _____

22. _____

20. 25)10,075

21. 18)10,944

22. 17)15,334

23. _____

24. _____

25. _____

26. _____

27. _____

28. _____

23. 18)16,218

24. 21)12,789

25. 19)17,195

29. _____

30. _____

31. _____

26. 39)12,051

27. 61)30,805

28. 55)16,940

29. 81)32,967

30. 72)57,888

31. 94)19,552

32. 25)50,125

33. 17)51,119

34. 23)69,046

ANSWERS

32. _____
33. _____
34. _____
35. _____
36. _____
37. _____
38. _____
39. _____
40. _____
41. _____
42. _____
43. _____
44. _____

35. 35)70,175

36. 38)228,342

37. 61)122,549

38. 52)416,208

39. 75)150,525

40. 225)901,350

41. 335)1,342,680

42. 724)2,899,690

43. 882)2,646,882

44. 741)4,452,669

Work Space

Applications

Name _____ Date_____

1. A word processing department can produce 1,632 letters in 8 hours. How many letters can they produce in one hour?

ANSWERS

1. _____

2. _____

3. _____

4. _____

2. A department store sold 525 leather coats during a special 5-day sale. What was the average number of coats sold daily?

3. A local school was given a grant of $83,200 to buy computers. How many computers can the school buy if the price of a computer is $800?

4. A family of 5 won $15,225 on a TV game show. What is each family member's share?

5. A large industrial machine can make 16,072 units in 8 hours. How many units can it make in one hour?

6. Sue Teal makes $12,600 a year. What does she make in a month?

7. Angela Kaye's new car gets an average of 35 miles to a gallon of gas. How many gallons of gas will it take her to drive 3,780 miles?

8. A gasoline tank holds 49,500 gallons of gas. How many days would it take to empty the tank if 550 gallons are taken out each day?

9. Tuna is packed 24 cans to a case. How many cases will be needed to pack 2,520 cans?

unit 2.5

Checking Division and Estimating Answers

After completing Unit 2.5, you will be able to:

1. Check the accuracy of division problems.
2. Estimate the answer to division problems.

To check for accuracy in a division problem, you multiply the quotient by the divisor. The result, if your math is correct, will equal the dividend.

Example

Divide 504 by 7 and check for accuracy.

Solution   **Accuracy Check**

$$
\begin{array}{r}
72 \\
7)\overline{504} \\
\underline{49} \\
14 \\
\underline{14} \\
0
\end{array}
\qquad
\begin{array}{r}
72 \;\longleftarrow \text{quotient} \\
\times \;\;7 \;\longleftarrow \text{divisor} \\
\hline
504
\end{array}
$$

If the problem has a remainder, add the remainder to the product of the quotient and the divisor.

Example

Divide 3,455 by 8 and check for accuracy.

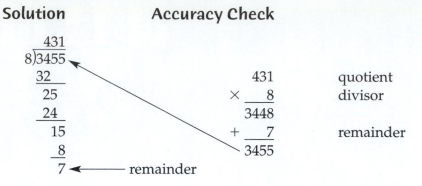

Solution	Accuracy Check	
	431	quotient
	× 8	divisor
	3448	
	+ 7	remainder
	3455	

remainder

Self-Check

Divide 3,864 by 24 and check your answer.

Solution

161

~~~ ESTIMATING ANSWERS IN DIVISION

We can estimate an answer in a division problem by rounding the divisor and the dividend to the value of its first digit, and then shortcut divide.

Example

Estimate the quotient of $46\overline{)8,120}$ and then find the actual quotient.

Solution

46 rounds to 50; 8,120 rounds to 8,000.

Estimate	Actual

$$\begin{array}{r} 160 \\ 5\cancel{0}\overline{)8,00\cancel{0}} \end{array} \qquad \begin{array}{r} 169 \\ 48\overline{)8,120} \\ \underline{48} \\ 332 \\ \underline{288} \\ 440 \\ \underline{432} \\ 8r \end{array}$$

Self-Check

Estimate the quotient of $58\overline{)5,940}$ and then find the actual quotient.

Solution

Estimate: 100
Actual: 102 r24

Name _____ **Date** _____

Exercises

Find the quotient in each of the following problems. State any remainder as a fraction. Check your answers by the method shown in this unit. Show the details of your work in the space provided.

Check | Check | Check
1. 53)3,810 | **2.** 28)5,688 | **3.** 81)9,922

1. _____

2. _____

3. _____

4. _____

5. _____

6. _____

Check | Check | Check
4. 76)21,845 | **5.** 25)250,300 | **6.** 5)599,999

Check   **Check**   **Check**

7. $47\overline{)3,710}$ 8. $78\overline{)15,866}$ 9. $39\overline{)1,914}$

Check

10. $895\overline{)396,211}$

Estimate the quotient in each of the following problems. Then find the actual quotient:

	Estimate	Actual			Estimate	Actual
11. $41\overline{)8,115}$				12. $29\overline{)91,250}$		

	Estimate	Actual			Estimate	Actual
13. $52\overline{)87,300}$				14. $205\overline{)522,640}$		

	Estimate	Actual			Estimate	Actual
15. $891\overline{)1,600,000}$				16. $28\overline{)301,200}$		

Name _____ Date _____

Applications

The sales report for Bravo Company for March follows:

	BRAVO COMPANY March Sales Report			
Item	Total Sales in Dollars	Number of Units Sold	=	Average Sales Price per Unit
A	$ 12,445	2,111		$_____
B	29,600	4,810		_____
C	31,235	5,990		_____
D	57,900	11,895		_____
E	139,347	10,124		_____
F	297,910	30,111		_____
G	176,500	5,754		_____

Estimate the average sales price per unit.

Finding Averages

After studying Unit 2.6, you will be able to:

1. Find an average of a group of numbers.

Many times in business, and in our personal lives, we may need to calculate an **average,** that is, a number that represents of a group of numbers. For example, your final grade in a course is usually based on an *average* of all your test scores. Businesses often calculate such things as average daily sales, average monthly sales per salesperson, average number of miles traveled per trip, average daily balance of customers' accounts, and so on.

To find the average, follow these steps:

Step 1: Add together all the numbers in a group.

Step 2: Divide this total by the number of numbers in the group.

Example 1

John Brandwein had these scores in a business math class: 90, 95, 80, 94, and 91. Find his average score.

Solution

Step 1: Find the total of all test scores:

$$90 + 95 + 80 + 94 + 91 = 450$$

Step 2: Divide the total of the test scores by the number of tests:

$$\begin{array}{r} 90 \\ 5\overline{)450} \end{array}$$ average score on five math tests

By the way, the average of a group of numbers is called the **mean** of the numbers. Thus, in Example 1, we could say that the *mean* score was 90. Now, before we move on to the end-of-unit exercises, let's look at an additional example.

Example 2

On Friday, six secretaries at Espon Supply Company typed 54 letters. What was the average number of letters typed by each secretary?

Solution

Divide 54 (the total number of letters typed) by 6 (the number of typists).

$$\frac{9}{6\overline{)54}} \quad \text{average number of letters typed}$$

Self-Check

Mary Boles is a real estate agent. During August she sold four houses. Her sales commissions on the four houses were $2,802, $3,200, $4,000, and $3,150. What was her average sales commission for the month?

Solution

$3,288

Name _____ **Date** _____

Exercises

Find the average of the following groups of numbers. State any remainder as a fraction:

1. 18, 12, 8, 19, 25

2. 2, 24, 5, 38, 4

3. 125, 235, 659, 120

4. 318, 295, 294, 269

5. 455, 699, 681, 965

6. 1,345, 5,682, 3,588

7. 1,890, 4,556, 5,690, 4,500

8. 12,515 and 28,900

9. 25,600 and 18,990

10. 55,675, 51,300, 87,000

ANSWERS

1. _____
2. _____
3. _____
4. _____
5. _____
6. _____
7. _____
8. _____
9. _____
10. _____

Find the average dollar amount of each of the following.

11. $50 + $80

12. $75 + $125

ANSWERS

11. _____

12. _____

13. _____

14. _____

15. _____

16. _____

17. _____

18. _____

13. $900 + $200

14. $875 + $325

15. $920 + $1,180

16. $125 + $300 + $375

17. $450 + $1,030 + $80 + $40

18. $12,850 + $5,150

Name _____ Date _____

Applications

1. Five sales associates from Dogwood Reality Company produced
    $1,200,000 in sales during the first quarter of the year. What was the
    average sales per associate?

ANSWERS

1. _____

2. _____

3. _____

4. _____

2. Bob Ziglar lost 37 pounds in 6 months. What was Bob's average weight
    loss for each month?

5. _____

3. The Baker family is planning to drive 2,321 miles on their vacation.
    How many miles will they have to average each day if they plan to
    make the trip in 3 days?

4. Louis Lobo is a salesperson in a department store. Last week he made
    these sales: Monday, $300; Tuesday, $320; Wednesday, $290; Thursday,
    $325; Friday, $250. What was the average of Louis's daily sales for the
    week?

5. The 1,200 workers at Harkon Industries built 72,000 units on Friday.
    How many units did each worker average?

6. Following are sales figures for Morez Company for the first two quarters of 20XX:

	First quarter			Second quarter		
Salesperson	Jan.	Feb.	March	April	May	June
Alvers	$12,000	$18,000	$14,500	$10,000	$8,000	$8,500
Bickers	6,000	13,500	5,500	12,400	6,600	4,000
Deal	22,000	28,000	14,000	14,500	11,000	9,000
Turley	12,300	9,700	8,000	8,200	6,000	12,800

Calculate the following:

a. average sales for January

b. average monthly sales for the first quarter

c. average monthly first-quarter sales for Deal

d. average sales for June

e. average sales for the second quarter

f. average monthly second-quarter sales for Bickers

g. average sales per salesperson for the first quarter

h. average sales per salesperson for the 6-month period

7. Katie McDermott earned the following sales commissions during her first 5 years as a textbook sales representative:

Year 1	Year 2	Year 3	Year 4	Year 5
$18,000	$12,000	$22,000	$28,000	$20,000

a. What were her average annual sales commissions for the 5-year period?

b. What annual sales commissions did she average in her first 3 years?

c. What annual sales commissions did she average in her last 2 years?

d. What monthly sales commissions did she average in her first year?

section

III

Working with Fractions

unit 3.1

The Meaning of Fractions

After completing Unit 3.1, you will be able to:

1. Identify the terms of a common fraction.
2. Differentiate among proper fractions, improper fractions, and mixed numbers.

Most of the numbers we have worked with so far have been whole numbers. We use whole numbers to represent complete units, such as a gallon of milk, a dollar, a car, a TV set, and a textbook.

There are many times where we need to express less than a complete unit. For example, you often buy a "half" gallon of milk, or you need two "quarters" to put into a soft drink machine, or you may plan to jog "three-quarters" of a mile today. In each of these examples, you are working with values that are less than one whole unit. In Unit 3.1, we will start our study of fractions, or partial units.

UNDERSTANDING FRACTIONS

A fraction is used to show that a whole unit has been divided into equal parts. Thus, a **fraction** is a *part of a whole unit*. For example, the following box represents one whole unit. (Remember that a whole unit is anything that can be counted or measured.) But, note that the box is divided into four equal parts and that three of the four parts are shaded. This means that we are identifying three of the four parts, or *three-fourths* of the whole box.

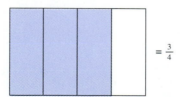

A fraction, or **common fraction,** is written by separating two numbers by a line, like this $\frac{3}{4}$.

The number that is written under the line shows how many equal parts the unit is divided into; it is called the **denominator.** The number that is written above the line is the **numerator;** it shows how many parts of the whole are expressed by the fraction. The line between the numerator and the denominator is called the **fraction bar.**

$\frac{3}{4}$ ◄——— **numerator** = the number of parts expressed by the fraction
◄——— **denominator** = the number of parts which the unit has been divided into

It is important to remember that the numerator always tells us the number of parts that are in the fraction. And, the denominator tells us how many parts into which the complete unit has been divided. So, the fraction $\frac{2}{3}$ tells us that a complete unit has been divided into three equal parts and that two of the three parts are represented by the fraction. The numerator and the denominator are called the **terms** of the fraction.

〰〰 TYPES OF COMMON FRACTIONS

There are three types of common fractions: proper fractions, improper fractions, and mixed numbers.

Proper fractions. A **proper fraction** is a fraction with a numerator that is less than its denominator. Examples include $\frac{1}{8}, \frac{1}{4}, \frac{1}{2}, \frac{3}{4}, \frac{7}{8}$, and $\frac{9}{10}$. A proper fraction expresses less than a whole unit.

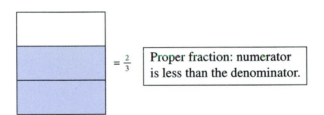

$= \frac{2}{3}$ Proper fraction: numerator is less than the denominator.

Improper fractions. An **improper fraction** is a fraction with a numerator that is equal to or greater than its denominator. Examples include $\frac{5}{5}, \frac{7}{6}, \frac{11}{9}, \frac{18}{5}, \frac{4}{3}$, and $\frac{19}{15}$. An improper fraction expresses one or more whole units.

$\frac{4}{4}$ $\qquad$ + $\qquad$ $\frac{1}{4}$

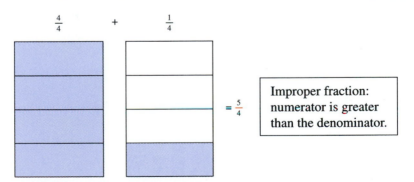

$= \frac{5}{4}$ Improper fraction: numerator is greater than the denominator.

Mixed numbers. A **mixed number** is a whole number and a common fraction. Examples include $3\frac{1}{2}, 4\frac{1}{8}$, and $75\frac{2}{3}$. A mixed number expresses a whole unit and a fraction.

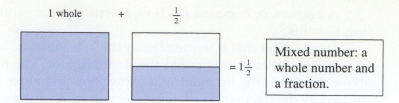

1 whole + $\frac{1}{2}$ $= 1\frac{1}{2}$

Mixed number: a whole number and a fraction.

Self-Check

Identify each of the following as either (1) a proper fraction, (2) an improper fraction, or (3) a mixed number:

(a) $\frac{1}{2}$ (b) $\frac{3}{9}$ (c) $\frac{5}{5}$ (d) $\frac{7}{5}$ (e) $4\frac{6}{7}$

Solution

(a) (1)   (b) (1)   (c) (2)   (d) (2)   (e) (3)

Exercises

Name _____ Date _____

Complete these statements:

1. A fraction is part of _____.

7. _____

2. A fraction is written by separating two numbers by _____.

8. _____

3. The top number in a fraction is called the _____.

12. _____

4. The bottom number in a fraction is called the _____.

5. What does the top number of a fraction tell you? _____
_____.

6. What does the bottom number of a fraction tell you? _____
_____.

7. In the fraction $\frac{6}{7}$, how many parts are being expressed by the fraction?

8. In the fraction $\frac{11}{12}$, into how many equal parts has the whole been divided?

9. A fraction whose top number is less than its bottom number is called a(n) _____ fraction.

10. A fraction whose top number is equal to or greater than its bottom number is called a(n) _____ fraction.

11. A _____ combines a whole number and a fraction.

12. If one complete unit is divided into six equal parts and two are expressed, what will be the denominator of the fraction?

13. Express each of the following as a fraction:

    a. three of four parts

    b. three of seven parts

    c. four of nine parts

    d. eleven of seventeen parts

    e. six of thirteen parts

    f. fifteen of thirty-one parts

    g. thirty-three of fifty-eight parts

    h. ninety-one of hundred parts

14. What fraction is shown by each of these pictures?

a.

b.

c.

d.

e.

f.

13. a. _____

    b. _____

    c. _____

    d. _____

    e. _____

    f. _____

    g. _____

    h. _____

14. a. _____

    b. _____

    c. _____

    d. _____

    e. _____

    f. _____

Work Space

unit 3.1

Applications

1. A high-school football team won 8 out of 12 football games. What fraction of their games did they win?

2. A business math class has 10 girls and 20 boys. What fraction of the class is made up of girls?

3. A quarterback completed 17 passes in 21 attempts. What fraction of the passes did he complete?

4. Out of a class of 45 students, 12 were sick. What fraction of the class was sick?

5. In May, Greg went on a diet to lose 15 pounds. By July, he had lost 11 pounds. What fraction of his weight loss goal has Greg achieved?

6. Twila bought a six-pack of cola to take to the beach. By 12 o'clock, Twila had finished four cans. What fraction of the cola is left for the afternoon?

7. Conrad bought a 32-ounce container of milk and used 8 ounces to make a milkshake. What fraction of the milk did Conrad use for the shake?

7. _____

8. _____

9. _____

10. _____

8. Out of 60 questions on a history test, a student marked 52 correct answers. What fraction of the test did the student get correct?

9. Kim's weekly paycheck is $225. Of this amount, Kim saves $75. What fraction of her paycheck does she save?

10. The price of a new car is $12,000, but you must make a down payment of $2,000. The down payment is what fraction of the price of the car?

Work Space

unit 3.2

Equivalent Fractions

After completing Unit 3.2, you will be able to:

1. Reduce fractions to lower terms.
2. Raise fractions to higher terms.
3. Find the greatest common divisor.

Let's say you ordered pizza when pizzas were on a "two for one" special. When the pizzas were delivered, you cut one into four equal parts and cut the second pizza into two equal parts. You now have two pizzas, each divided into equal parts. One pizza has four equal parts and the other pizza has two equal parts. If you ate two pieces of the pizza cut into four parts, you would have eaten $\frac{2}{4}$ of the pizza. But, if you ate one piece of the pizza cut into two parts, you would have eaten $\frac{1}{2}$ of the pizza.

Either way, you would have eaten the same amount of pizza because 2 parts out of 4 is the same amount as 1 part out of 2. In other words, $\frac{2}{4} = \frac{1}{2}$. Fractions that represent the same value are called **equivalent fractions.**

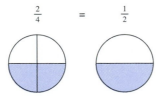

$$\frac{2}{4} \quad = \quad \frac{1}{2}$$

$\frac{2}{4}$ and $\frac{1}{2}$ are equivalent fractions because they have the same value.

≈ REDUCING FRACTIONS TO LOWEST TERMS

Fractions with smaller terms—remember the *terms* of a fraction are the numerator and the denominator—are easier to work with than fractions with larger terms. So, answers to problems dealing with fractions should be reduced to **lowest terms.** To reduce a fraction to its lowest terms, select any number (other than 1) that will divide evenly into *both* the numerator and the denominator. Keep dividing until the terms can no longer be evenly divided by the same number. The following rule will help you when reducing fractions:

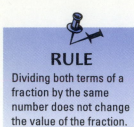

RULE

Dividing both terms of a fraction by the same number does not change the value of the fraction.

Example 1

Reduce $\frac{6}{8}$ to its lowest terms.

Solution

Select a number that will divide evenly into both the numerator and the denominator. By simple observation, we can see that 2 will divide evenly into both terms of this fraction.

$$\frac{6}{8} = \frac{6 \div 2}{8 \div 2} = \frac{3}{4}$$

We have now reduced $\frac{6}{8}$ to $\frac{3}{4}$. The fraction is now in its lowest terms because there is no number that can be divided evenly into both the 3 and the 4.

Example 2

Reduce $\frac{25}{75}$ to its lowest terms.

Solution

Select a number that will divide evenly into the terms of the fraction. Let's use 5.

$$\frac{25}{75} = \frac{25 \div 5}{75 \div 5} = \frac{5}{15}$$

We have now reduced $\frac{25}{75}$ to $\frac{5}{15}$. But, is the fraction in its lowest terms? No. Because $\frac{5}{15}$ can be reduced again. Let's finish reducing it by again dividing both terms by 5:

$$\frac{5}{15} = \frac{5 \div 5}{15 \div 5} = \frac{1}{3}$$

Not all fractions can be reduced. If there is no number that will divide evenly into both the numerator and the denominator, the fraction is already in its lowest terms.

Example

Reduce $\frac{13}{17}$ to its lowest terms.

Solution

There is no number that will divide evenly into both the numerator and the denominator. Thus, the terms of the fraction cannot be reduced.

Self-Check

Where possible, reduce these fractions to its lowest terms:

(a) $\frac{4}{8}$ (b) $\frac{8}{24}$ (c) $\frac{8}{32}$ (d) $\frac{11}{13}$ (e) $\frac{42}{80}$

Solution

(a) $\frac{1}{2}$ (b) $\frac{1}{3}$ (c) $\frac{1}{4}$ (d) $\frac{11}{13}$ (e) $\frac{21}{40}$

DETERMINING THE GREATEST COMMON DIVISOR

In the previous section, we learned that a fraction is reduced to its *lowest terms* by dividing its terms by the same number until further division is not possible. In the examples used, we selected a number that would divide evenly into both terms and continued to divide until the terms could no longer be divided evenly by the same number. Basically, this is a method of trial and error. We select a number and determine if it will divide evenly into both terms of the fraction. If it does not, we select another number and try again. This method works well for many fractions. However, it can be time-consuming if numbers that are common to the numerator and the denominator are not obvious.

For example, how would you reduce the fraction $\frac{39}{65}$? You cannot use 3 because only the numerator (39) is evenly divisible by 3; the denominator (65) is not. You cannot use 2 because neither term is evenly divisible by 2. So, what number do you use?

You can use the following method to determine the **greatest common divisor (GCD),** that is, the largest number that will divide evenly into both the numerator and the denominator.

 Step 1: Divide the numerator into the denominator. If it divides evenly without a remainder, the numerator is the greatest common divisor. If a remainder exists, go on to Step 2.

Step 2: Divide the remainder into the divisor from Step 1.

 Step 3: Keep dividing the remainders into the previous divisors until there is no remainder. The *last divisor* will be your greatest common divisor.

Note: If you obtain a remainder of 1, it means that the fraction is already in its lowest terms.

Now let's find the greatest common divisor and reduce $\frac{39}{65}$ to its lowest terms:

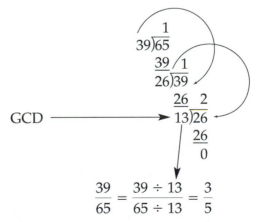

$$\frac{39}{65} = \frac{39 \div 13}{65 \div 13} = \frac{3}{5}$$

Even though you can use any common divisor to reduce a fraction, you should always use the greatest common divisor because the GCD will reduce the fraction to its lowest terms in one step.

≋ RAISING FRACTIONS TO HIGHER TERMS (HIGHER EQUIVALENTS)

RULE

Multiplying both terms of a fraction by the same number does not change the value of the fraction.

NOTE

In raising this fraction to higher terms, we multiplied both terms by 3. Because $\frac{3}{3} = 1$, multiplying or dividing by such a number does not change the value of the fraction because the fraction is actually being multiplied or divided by 1.

Although final answers dealing with fractions are usually reduced to lowest terms, there are some operations with fractions in which you will need to *raise* a fraction to higher terms. For example, to add and subtract fractions with different denominators, you usually need to raise a fraction to higher terms.

A fraction is raised to higher terms by multiplying *both* terms of the fraction by the *same* number. The rule to the left will help you when raising fractions to higher terms:

For example, we can raise $\frac{1}{2}$ to higher terms by multiplying both the numerator and the denominator by any number we choose. Let's use 3 for this example:

$$\frac{1}{2} = \frac{1 \times 3}{2 \times 3} = \frac{3}{6}$$

We have now raised $\frac{1}{2}$ to $\frac{3}{6}$, which is a higher equivalent. In effect, we have just renamed the fraction because $\frac{1}{2}$ and $\frac{3}{6}$ have the same value; they are equivalent.

$\frac{1}{2}$ and $\frac{3}{6}$ are equivalent fractions because they represent the same value.

≋ RAISING A FRACTION TO HIGHER TERMS WHEN A NEW DENOMINATOR IS ALREADY GIVEN

Sometimes you will need to raise a fraction to higher terms when a new denominator (the higher term) is already given. For example,

$$\frac{3}{4} = \frac{?}{16}$$

To raise such a fraction to higher terms, follow these steps:

Step 1: Divide the original denominator into the new denominator.

Step 2: Multiply the numerator by the quotient you found in Step 1, and write the result over the new denominator.

To continue the example, let's now raise $\frac{3}{4}$ to a higher equivalent with a denominator of 16.

Step 1: Divide the original denominator, 4, into the new denominator, 16:

$$\frac{3}{4} = \frac{?}{16} \longrightarrow 4\overline{)16}^{4}$$

Step 2: Multiply the numerator by the 4 you obtained in Step 1 ($4 \times 3 = 12$) and place the result over the new denominator:

$$\boxed{4 \times 3 = 12} \longrightarrow \quad \frac{3}{4} = \frac{12}{16}$$

We know that our answer is correct because:

$$\frac{4}{4} \times \frac{3}{4} = \frac{12}{16}$$

> Remember that the value of the fraction is not changed; we only renamed the fraction because we multiplied by $\frac{4}{4}$, which is equal to 1.

Self-Check

Raise these fractions to the term indicated:

(a) $\dfrac{2}{3} = \dfrac{?}{12}$ (b) $\dfrac{3}{16} = \dfrac{?}{32}$ (c) $\dfrac{4}{7} = \dfrac{?}{28}$

Solution

(a) $\dfrac{8}{12}$ (b) $\dfrac{6}{32}$ (c) $\dfrac{16}{28}$

Name _____ Date_____

Exercises

Where possible, reduce each of these fractions to its lowest terms:

ANSWERS

1. $\dfrac{6}{12}$ 2. $\dfrac{4}{8}$ 3. $\dfrac{3}{4}$ 4. $\dfrac{6}{9}$ 5. $\dfrac{5}{8}$

6. $\dfrac{9}{12}$ 7. $\dfrac{4}{16}$ 8. $\dfrac{8}{12}$ 9. $\dfrac{11}{12}$ 10. $\dfrac{16}{24}$

11. $\dfrac{8}{16}$ 12. $\dfrac{7}{14}$ 13. $\dfrac{21}{33}$ 14. $\dfrac{18}{45}$ 15. $\dfrac{12}{36}$

16. $\dfrac{22}{26}$ 17. $\dfrac{34}{66}$ 18. $\dfrac{12}{80}$ 19. $\dfrac{72}{81}$ 20. $\dfrac{14}{28}$

21. $\dfrac{56}{92}$ 22. $\dfrac{44}{80}$ 23. $\dfrac{36}{63}$ 24. $\dfrac{45}{50}$ 25. $\dfrac{10}{50}$

1. _____
2. _____
3. _____
4. _____
5. _____
6. _____
7. _____
8. _____
9. _____
10. _____
11. _____
12. _____
13. _____
14. _____
15. _____
16. _____
17. _____
18. _____
19. _____
20. _____
21. _____
22. _____
23. _____
24. _____
25. _____

26. $\dfrac{40}{90}$ **27.** $\dfrac{45}{90}$ **28.** $\dfrac{35}{45}$ **29.** $\dfrac{21}{63}$ **30.** $\dfrac{16}{54}$

31. $\dfrac{22}{44}$ **32.** $\dfrac{25}{75}$ **33.** $\dfrac{36}{36}$ **34.** $\dfrac{72}{96}$ **35.** $\dfrac{90}{95}$

36. $\dfrac{120}{240}$ **37.** $\dfrac{50}{150}$ **38.** $\dfrac{75}{300}$ **39.** $\dfrac{280}{340}$

Find the greatest common divisor and reduce these fractions to its lowest terms.

40. $\dfrac{35}{75}$ **41.** $\dfrac{45}{135}$ **42.** $\dfrac{30}{92}$ **43.** $\dfrac{75}{225}$ **44.** $\dfrac{68}{238}$

45. $\dfrac{72}{192}$ **46.** $\dfrac{22}{324}$ **47.** $\dfrac{55}{255}$ **48.** $\dfrac{234}{288}$ **49.** $\dfrac{350}{725}$

50. $\dfrac{675}{900}$ **51.** $\dfrac{250}{550}$ **52.** $\dfrac{430}{920}$ **53.** $\dfrac{750}{1,000}$ **54.** $\dfrac{540}{1,000}$

55. $\dfrac{270}{1,200}$ **56.** $\dfrac{250}{1,500}$ **57.** $\dfrac{300}{1,200}$ **58.** $\dfrac{525}{2,500}$ **59.** $\dfrac{450}{1,600}$

ANSWERS

26. _____
27. _____
28. _____
29. _____
30. _____
31. _____
32. _____
33. _____
34. _____
35. _____
36. _____
37. _____
38. _____
39. _____
40. _____
41. _____
42. _____
43. _____
44. _____
45. _____
46. _____
47. _____
48. _____
49. _____
50. _____
51. _____
52. _____
53. _____
54. _____
55. _____
56. _____
57. _____
58. _____
59. _____

Raise these fractions to the term indicated:

60. $\dfrac{3}{4} = \dfrac{?}{8}$

61. $\dfrac{2}{3} = \dfrac{?}{21}$

62. $\dfrac{3}{5} = \dfrac{?}{15}$

63. $\dfrac{2}{7} = \dfrac{?}{28}$

64. $\dfrac{4}{5} = \dfrac{?}{20}$

65. $\dfrac{7}{9} = \dfrac{?}{45}$

66. $\dfrac{5}{8} = \dfrac{?}{40}$

67. $\dfrac{3}{5} = \dfrac{?}{15}$

68. $\dfrac{5}{6} = \dfrac{?}{30}$

69. $\dfrac{3}{7} = \dfrac{?}{21}$

70. $\dfrac{7}{8} = \dfrac{?}{40}$

71. $\dfrac{2}{3} = \dfrac{?}{24}$

72. $\dfrac{8}{9} = \dfrac{?}{72}$

73. $\dfrac{7}{8} = \dfrac{?}{64}$

74. $\dfrac{2}{7} = \dfrac{?}{28}$

75. $\dfrac{8}{12} = \dfrac{?}{36}$

Raise each of these fractions as indicated:

76. $\dfrac{1}{4}$ to 16ths

77. $\dfrac{3}{8}$ to 32nds

78. $\dfrac{2}{15}$ to 45ths

79. $\dfrac{5}{9}$ to 72nds

80. $\dfrac{9}{12}$ to 48ths

81. $\dfrac{14}{15}$ to 75ths

82. $\dfrac{32}{40}$ to 80ths

83. $\dfrac{17}{32}$ to 64ths

84. $\dfrac{12}{18}$ to 36ths

60. _____

61. _____

62. _____

63. _____

64. _____

65. _____

66. _____

67. _____

68. _____

69. _____

70. _____

71. _____

72. _____

73. _____

74. _____

75. _____

76. _____

77. _____

78. _____

79. _____

80. _____

81. _____

82. _____

83. _____

84. _____

Applications

Name _____ Date_____

In these problems, reduce each final answer to its lowest terms, if needed:

ANSWERS

1. Maureen O'Shea can jog 4 miles. Her goal is to be able to jog 6 miles. What fraction of her goal has she reached?

1. _____

2. _____

3. _____

2. Bill Seymour's gross pay was $200, but $40 was taken out for taxes. What fraction of Bill's gross pay was taken out for taxes?

4. _____

5. _____

6. a. _____

3. In a clearance sale, a department store sold 50 fur coats. What fraction of fur coats was sold if the original number in stock was 70?

b. _____

c. _____

7. _____

4. Of a graduating class of 600, there were 350 girls. What fraction of the class was girls?

5. Wendy Rosenburg jogged 20 minutes out of an hour she spent on a treadmill. (a) What fraction of the hour did she spend jogging? (b) What fraction of the hour did she spend walking?

6. What fraction of an hour is:

   a. 15 minutes

   b. 30 minutes

   c. 45 minutes

7. Terry Applebee, a salesperson, visited 36 customers and made sales to 12 of them. With what fraction of her customers was she successful?

8. A new house is for sale at $90,000 with a required down payment of $18,000. The down payment is what fraction of the price of the house?

9. Out of a total graduating class of 650 students, 525 went on to college. What fraction of the graduating class *did not* go on to college.

10. During a clearance sale, Windsor Men's Shop sold 148 blazers. What fraction of the total stock was sold if there were 240 blazers in the store before the sale began?

11. Using equivalent fractions, change each of these amounts as indicated:

a. $\frac{3}{4}$ of a quart to ounces (32nds)

b. $\frac{1}{8}$ of a pound to ounces (16ths)

8. _____

9. _____

10. _____

11. a. _____

b. _____

unit 3.3

Improper Fractions and Mixed Numbers

After completing Unit 3.3, you will be able to:

1. Identify improper fractions and mixed numbers.
2. Convert mixed numbers to improper fractions.
3. Convert improper fractions to whole or mixed numbers.

You know that a fraction is used to express a part of a whole unit, or something less than one. There are times, however, when fractions are used to express more than a unit. For example, you know that a quarter is $\frac{1}{4}$ of a dollar (a quarter can be written as $\$\frac{1}{4}$). If you had 3 quarters, you would have $\frac{3}{4}$ of a dollar. But, what if you had 5 quarters? Then you would have $\frac{5}{4}$ of a dollar, which means that you have one dollar ($\frac{4}{4}$) plus $\frac{1}{4}$ of another dollar.

$$\frac{4}{4} \quad \text{and} \quad \frac{1}{4} \quad = \frac{5}{4}$$

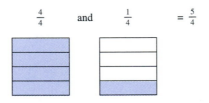

We call $\frac{5}{4}$ improper because its numerator is larger than its denominator. (Remember that an improper fraction is one whose numerator is equal to or greater than its denominator.) Other examples of improper fractions are $\frac{6}{5}, \frac{8}{8}, \frac{11}{9}, \frac{21}{10}$, and $\frac{75}{5}$.

≋ MIXED NUMBERS

A mixed number is made up of a whole number and a fraction. Examples of mixed numbers include $2\frac{1}{2}$, $3\frac{1}{3}$, $5\frac{3}{4}$, and $8\frac{6}{7}$. A mixed number represents a value that is equal to or greater than one whole unit. If you lost $5\frac{1}{2}$ pounds, for example, you would have lost five whole pounds, plus $\frac{1}{2}$ of another pound.

Mixed numbers can be expressed as improper fractions, and all improper fractions can be expressed as mixed numbers. Let's look at how it is done.

≋ CONVERTING MIXED NUMBERS TO IMPROPER FRACTIONS

To convert (to change) a mixed number to an improper fraction, follow these steps:

Step 1: Multiply the whole number by the denominator of the fraction.

Step 2: Add the product you obtained in Step 1 to the numerator of the fraction.

Step 3: Write the total from Step 2 over the original denominator.

Example 1

Convert $6\frac{1}{3}$ to an improper fraction.

Solution

Step 1: Multiply the whole number, 6, by the denominator of the fraction, 3:

$$6\frac{1}{3} \rightarrow 3 \times 6 = 18$$

Step 2: Add the product from Step 1 (18) to the numerator of the fraction (1):

$$18 + 1 = 19$$

Step 3: Write the total from Step 2 (19) over the original denominator:

$$6\frac{1}{3} = (3 \times 6 = 18) + 1 = \frac{19}{3}$$

Example 2

Convert $6\frac{7}{8}$ to an improper fraction.

Solution

$$6\frac{7}{8} = (8 \times 6 = 48) + 7 = \frac{55}{8}$$

CONVERTING IMPROPER FRACTIONS TO WHOLE OR MIXED NUMBERS

To convert an improper fraction to a whole or mixed number, you divide the denominator into the numerator. If there is no remainder, the answer is a whole number. If there is a remainder, the improper fraction is a mixed number; state the remainder as a fraction over the original denominator.

Example 1

Convert $\frac{45}{9}$ to a whole or mixed number.

Solution

$$\begin{array}{r} 5 \\ 9\overline{)45} \end{array} \qquad = 5$$

> Since the problem came out even (no remainder), the answer is a whole number.

Example 2

Convert $\frac{21}{5}$ to a whole or mixed number.

Solution

$$\begin{array}{r} 4 \\ 5\overline{)21} \\ \underline{20} \\ \text{remainder} \longrightarrow 1 \end{array} \qquad = 4\frac{1}{5}$$

> Since the problem had a remainder, the answer is a mixed number.

Example 3

Convert $\frac{55}{8}$ to a whole or mixed number:

Solution

$$\begin{array}{r} 6 \\ 8\overline{)55} \\ \underline{48} \\ 7 \end{array} \qquad = 6\frac{7}{8}$$

Self-Check

Convert $\frac{51}{12}$ to a whole or mixed number. Reduce the fractional part of the answer to its lowest terms.

Solution

$4\frac{1}{4}$

unit 3.3

Exercises

Change each of these mixed numbers to improper fractions:

1. $3\frac{3}{4}$ **2.** $5\frac{1}{4}$ **3.** $6\frac{7}{8}$ **4.** $12\frac{1}{8}$ **5.** $15\frac{1}{3}$ **6.** $22\frac{1}{2}$ **7.** $9\frac{2}{3}$

8. $15\frac{1}{2}$ **9.** $6\frac{1}{15}$ **10.** $7\frac{4}{5}$ **11.** $10\frac{1}{4}$ **12.** $21\frac{2}{3}$ **13.** $5\frac{1}{9}$ **14.** $3\frac{5}{6}$

15. $14\frac{5}{6}$ **16.** $16\frac{4}{5}$

Change each of these improper fractions to a mixed number or to a whole number. Reduce the fractional part of your answers:

17. $\frac{12}{5}$ **18.** $\frac{13}{7}$ **19.** $\frac{34}{6}$ **20.** $\frac{11}{8}$ **21.** $\frac{9}{2}$ **22.** $\frac{15}{3}$ **23.** $\frac{17}{4}$

ANSWERS

1. _____
2. _____
3. _____
4. _____
5. _____
6. _____
7. _____
8. _____
9. _____
10. _____
11. _____
12. _____
13. _____
14. _____
15. _____
16. _____
17. _____
18. _____
19. _____
20. _____
21. _____
22. _____
23. _____

24. $\dfrac{17}{8}$ **25.** $\dfrac{39}{5}$ **26.** $\dfrac{31}{7}$ **27.** $\dfrac{25}{3}$ **28.** $\dfrac{47}{4}$ **29.** $\dfrac{52}{7}$ **30.** $\dfrac{15}{2}$

ANSWERS

24. _____

25. _____

26. _____

27. _____

28. _____

29. _____

30. _____

unit
3.3

Applications

1. How much money does each of these improper fractions represent?

   a. $\dfrac{\$12}{4} =$

   b. $\dfrac{\$15}{5} =$

   c. $\dfrac{\$21}{5} =$

   d. $\dfrac{\$18}{6} =$

   e. $\dfrac{\$33}{10} =$

ANSWERS

1. a. _____
   b. _____
   c. _____
   d. _____
   e. _____
2. _____
3. _____
4. _____

2. Write 37 months in terms of years.

3. Write 12 quarters in terms of dollars. (*Hint:* One quarter equals $\frac{1}{4}$ of a dollar.)

4. A play was on Broadway for 32 months. How many years was it on Broadway?

5. Write the following mixed numbers as improper fractions:

    a. $12\frac{1}{4}$

    b. $6\frac{7}{8}$

    c. $5\frac{3}{4}$

    d. $14\frac{1}{3}$

5. **a.** _____

    b. _____

    c. _____

    d. _____

Work Space

unit 3.4

Adding Fractions and Mixed Numbers

After completing Unit 3.4, you will be able to:

1. Add fractions with the same denominator.
2. Add fractions with different denominators.
3. Add mixed numbers.

Fractions must have the same denominator to be added. If you are working with two or more fractions that already have the same denominator (or **common denominator**), the addition process is very simple: Add the numerators and place the sum over the common denominator; reduce the final answer to its lowest terms, if needed.

Example

Add these fractions $\frac{2}{15} + \frac{7}{15}$.

Solution

Step 1: Add the numerators:

$$2 + 7 = 9$$

Step 2: Place the sum of the numerators, 9, over the common denominator, 15:

$$\frac{2}{15} + \frac{7}{15} = \frac{9}{15}$$

Step 3: Reduce the answer to its lowest terms. (Remember, to reduce a fraction, find a number that will divide evenly into both the numerator and the denominator; the number needed to reduce this fraction is 3.)

$$\frac{9 \div 3}{15 \div 3} = \frac{3}{5}$$

If you add a set of fractions and get an improper fraction as your sum, you should convert it to a whole number or a mixed number.

Example
Add: $\frac{3}{5} + \frac{2}{5} + \frac{4}{5}$.

Solution
Step 1: Add the numerators:

$$3 + 2 + 4 = 9$$

Step 2: Place the sum of the numerators, 9, over the common denominator, 5:

$$\frac{3}{5} + \frac{2}{5} + \frac{4}{5} = \frac{9}{5}$$

Step 3: Convert the improper fraction to a mixed number:

$$\frac{9}{5} \quad = \quad 5\overline{)9} \quad = \quad 1\frac{4}{5}$$
$$\phantom{\frac{9}{5} \quad = \quad} \underline{5}$$
$$\phantom{\frac{9}{5} \quad = \quad} 4$$

REMEMBER

When adding fractions with the same denominator, *add only the numerators.*

Self-Check

Add these fractions. Change improper fractions to whole or mixed numbers and reduce all final answers to their lowest terms.

(a) $\frac{1}{3} + \frac{2}{3}$ **(b)** $\frac{3}{8} + \frac{5}{8}$ **(c)** $\frac{1}{5} + \frac{3}{5}$ **(d)** $\frac{3}{7} + \frac{4}{7} + \frac{5}{7}$

Solution

(a) 1   **(b)** 1   **(c)** $\frac{4}{5}$ **(d)** $1\frac{5}{7}$

ADDING FRACTIONS WITH DIFFERENT DENOMINATORS

In many operations dealing with fractions, you will have to add fractions that do not have the same denominator. To add a group of fractions with *unlike* denominators, you must change each fraction to an equivalent fraction with a common denominator.

To find a common denominator, you find a number into which all denominators will evenly divide. This can often be done by simple observation.

Example
Add: $\frac{1}{2} + \frac{1}{3}$.

Solution
Since halves and thirds are not common terms, we must find a common denominator. By simple observation, we can see that both

denominators will divide evenly into 6. Thus, 6 is a common denominator. We now raise each fraction to 6ths and add the numerators:

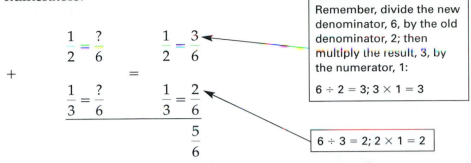

Remember, divide the new denominator, 6, by the old denominator, 2; then multiply the result, 3, by the numerator, 1:

$6 \div 2 = 3; 3 \times 1 = 3$

$\dfrac{1}{2} = \dfrac{?}{6}$ $\dfrac{1}{2} = \dfrac{3}{6}$

$+$

$\dfrac{1}{3} = \dfrac{?}{6}$ $\dfrac{1}{3} = \dfrac{2}{6}$

$= \dfrac{5}{6}$

$6 \div 3 = 2; 2 \times 1 = 2$

Since a set of fractions can have many common denominators, you can make the problem easier by using the **least common denominator (LCD).** The least common denominator is the smallest number into which all denominators will evenly divide. In the preceding example, for instance, the LCD is 6 because 6 is the smallest number into which both 2 and 3 will divide evenly.

FINDING THE LCD USING THE PRIME NUMBER METHOD

When a common denominator cannot be determined by simple observation, you can find the LCD by using the **prime number method.** A **prime number** is a number that can be divided evenly by itself and one, but no other number. Examples of prime numbers are 2, 3, 5, 7, 11, 13, 17, 19, and 23. To illustrate how to find the LCD using the prime number method, let's look at the next example.

Example
Find the LCD and add: $\frac{1}{6} + \frac{5}{12} + \frac{4}{15}$.

Solution
Step 1: Write the denominators in a row, like this:

$$)6 - 12 - 15$$

Step 2: Choose a prime number that will divide evenly into as many denominators as possible. For this example, let's start with 3:

$$\begin{array}{r} 3)6 - 12 - 15 \\ \hline 2 - 4 - 5 \end{array}$$

Step 3: Choose another prime number that will divide evenly into as many numbers as possible in the second row. Let's use 2 because it will divide evenly into the 2 and the 4:

$$\begin{array}{r} 3)6 - 12 - 15 \\ 2)2 - 4 - 5 \\ \hline 1 - 2 - 5 \end{array}$$

Note: Since 2 does not divide evenly into 5, simply bring down the 5.

Step 4: Continue dividing by a prime number until only 1's are left as quotients:

$$
\begin{array}{r}
3)\underline{6 - 12 - 15} \\
2)\underline{2 - 4 - 5} \\
5)\underline{1 - 2 - 5} \\
2)\underline{1 - 2 - 1} \\
1 - 1 - 1
\end{array}
$$

Step 5: Multiply all prime number divisors; the product is the LCD:

$$3 \times 2 \times 5 \times 2 = 60 \longleftarrow \text{LCD}$$

Now that we found the LCD of 60, we can finish the problem by raising each fraction to 60ths, and adding the numerators:

$$
\begin{array}{r}
\dfrac{1}{6} = \dfrac{10}{60} \\[2mm]
+ \dfrac{5}{12} = \dfrac{25}{60} \\[2mm]
\dfrac{4}{15} = \dfrac{16}{60} \\[2mm]
\dfrac{51}{60} = \dfrac{51 \div 3}{60 \div 3} = \dfrac{17}{20}
\end{array}
$$

Trial + error would be faster than this method, I think...

REMEMBER

Fractions with unlike terms must be converted to equivalent fractions before they can be added.

Self-Check

Find a common denominator and add the following fractions:

(a) $\dfrac{3}{4}$

$+ \dfrac{1}{6}$

(b) $\dfrac{5}{7}$

$+ \dfrac{2}{9}$

(c) $\dfrac{7}{8}$

$+ \dfrac{3}{4}$

Solution

(a) $\dfrac{11}{12}$

(b) $\dfrac{59}{63}$

(c) $1\dfrac{5}{8}$

ADDING MIXED NUMBERS

As you remember, a mixed number is a whole number and a fraction. Thus, when adding mixed numbers, you should work in separate steps, as follows:
To add mixed numbers:

Step 1: Add the fractions.

Step 2: Add the whole numbers.

Step 3: Combine the totals from Steps 1 and 2.

Example

Add $10\frac{1}{2} + 8\frac{2}{5}$.

Solution

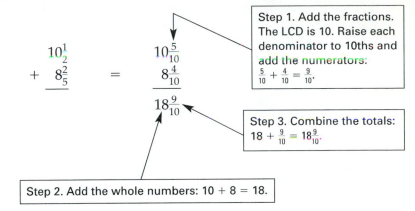

$$10\frac{1}{2}$$
$$+ \quad 8\frac{2}{5}$$

$=$

$$10\frac{5}{10}$$
$$8\frac{4}{10}$$
$$18\frac{9}{10}$$

Step 1. Add the fractions. The LCD is 10. Raise each denominator to 10ths and add the numerators: $\frac{5}{10} + \frac{4}{10} = \frac{9}{10}$.

Step 3. Combine the totals: $18 + \frac{9}{10} = 18\frac{9}{10}$.

Step 2. Add the whole numbers: $10 + 8 = 18$.

When you add mixed numbers, the sum of the fractions may be an improper fraction. If this happens, convert the improper fraction to a mixed number, then add the whole numbers.

Example

Add: $12\frac{2}{3} + 8\frac{3}{4}$.

Solution

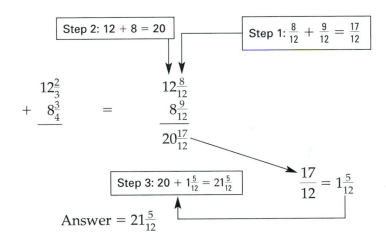

Step 2: $12 + 8 = 20$

Step 1: $\frac{8}{12} + \frac{9}{12} = \frac{17}{12}$

$$12\frac{2}{3}$$
$$+ \quad 8\frac{3}{4}$$

$=$

$$12\frac{8}{12}$$
$$8\frac{9}{12}$$
$$20\frac{17}{12}$$

$$\frac{17}{12} = 1\frac{5}{12}$$

Step 3: $20 + 1\frac{5}{12} = 21\frac{5}{12}$

$$\text{Answer} = 21\frac{5}{12}$$

Notice that the improper fraction $\frac{17}{12}$ was changed to a mixed number, $1\frac{5}{12}$; the mixed number was then added to the total of the whole numbers: $20 + 1\frac{5}{12} = 21\frac{5}{12}$.

Self-Check

Add the following mixed numbers:

$8\frac{3}{4} + 7\frac{1}{2}$.

Solution

$16\frac{1}{4}$

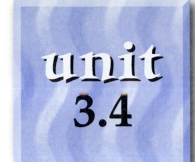

Exercises

Name _____ Date_____

Add these fractions. Change improper fractions to whole or mixed numbers and reduce all final answers to their lowest terms:

1. $\dfrac{1}{3} + \dfrac{2}{3}$

2. $\dfrac{3}{8} + \dfrac{5}{8}$

3. $\dfrac{1}{5} + \dfrac{3}{5}$

4. $\dfrac{2}{9} + \dfrac{6}{9}$

5. $\dfrac{3}{7} + \dfrac{4}{7} + \dfrac{5}{7}$

6. $\dfrac{1}{8} + \dfrac{3}{8} + \dfrac{7}{8}$

7. $\dfrac{2}{15} + \dfrac{4}{15} + \dfrac{7}{15}$

8. $\dfrac{3}{11} + \dfrac{4}{11} + \dfrac{6}{11}$

1. _____
2. _____
3. _____
4. _____
5. _____
6. _____
7. _____
8. _____

9. $\dfrac{5}{12} + \dfrac{7}{12} + \dfrac{11}{12}$

10. $\dfrac{3}{16} + \dfrac{9}{16} + \dfrac{1}{16}$

9. _____

10. _____

11. _____

12. _____

13. _____

14. _____

15. _____

16. _____

17. _____

18. _____

11. $\dfrac{8}{25} + \dfrac{11}{25} + \dfrac{16}{25}$

12. $\dfrac{5}{21} + \dfrac{8}{21} + \dfrac{11}{21}$

Add these fractions. Change improper fractions to whole numbers or mixed numbers and reduce all final answers to their lowest terms. (*Hint:* If both denominators are prime, just multiply them together to get the least common denominator.)

13. $\dfrac{3}{4}$

$+\dfrac{1}{6}$

14. $\dfrac{5}{7}$

$+\dfrac{2}{9}$

15. $\dfrac{7}{8}$

$+\dfrac{3}{4}$

16. $\dfrac{2}{5}$

$+\dfrac{1}{3}$

17. $\dfrac{9}{11}$

$+\dfrac{3}{5}$

18. $\dfrac{1}{4}$

$+\dfrac{2}{3}$

19. $\dfrac{2}{15}$
$+\ \dfrac{7}{30}$

20. $\dfrac{13}{15}$
$+\ \dfrac{11}{45}$

21. $\dfrac{7}{25}$
$+\ \dfrac{12}{30}$

ANSWERS

19. _____

20. _____

21. _____

22. _____

23. _____

24. _____

25. _____

26. _____

22. $\dfrac{5}{6}$
$\dfrac{2}{3}$
$+\ \dfrac{3}{4}$

23. $\dfrac{7}{8}$
$\dfrac{3}{8}$
$+\ \dfrac{2}{9}$

24. $\dfrac{12}{17}$
$\dfrac{12}{17}$
$\dfrac{11}{34}$

25. $\dfrac{21}{25}$
$\dfrac{8}{15}$
$+\ \dfrac{13}{20}$

26. $\dfrac{3}{7}$
$\dfrac{2}{21}$
$+\ \dfrac{11}{42}$

Add these mixed numbers. Reduce final answers to their lowest terms:

27. $8\frac{3}{4}$
$7\frac{1}{2}$

28. $5\frac{7}{8}$
$4\frac{4}{5}$

29. $8\frac{1}{3}$
$3\frac{2}{7}$

30. $9\frac{3}{5}$
$3\frac{1}{6}$

31. $3\frac{4}{7}$
$2\frac{3}{8}$

32. $7\frac{1}{4}$
$5\frac{2}{9}$

33. $9\frac{3}{8}$
$7\frac{5}{7}$

34. $3\frac{2}{5}$
$2\frac{1}{9}$

35. $15\frac{2}{9}$
$5\frac{3}{9}$
$17\frac{4}{5}$

36. $6\frac{7}{8}$
$25\frac{1}{5}$
$24\frac{3}{4}$

37. $23\frac{5}{9}$
$37\frac{8}{9}$
$55\frac{7}{8}$

38. $15\frac{14}{15}$
$14\frac{12}{25}$
$21\frac{13}{50}$

39. $22\frac{11}{14}$
$15\frac{9}{56}$
$7\frac{15}{28}$

40. $19\frac{11}{16}$
$9\frac{3}{4}$
$17\frac{13}{48}$

27. _____
28. _____
29. _____
30. _____
31. _____
32. _____
33. _____
34. _____
35. _____
36. _____
37. _____
38. _____
39. _____
40. _____

Applications

Name _____ Date_____

1. A large pizza was cut into 8 pieces. Bill ate 2 pieces, Jane ate 2 pieces, Sue at 1 piece, and Tim ate 2 pieces. What fraction of the pizza was eaten?

2. A recipe for bran muffins calls for $\frac{1}{2}$ cup whole wheat flour, $\frac{1}{2}$ cup wheat bran, $\frac{1}{2}$ cup honey, and $\frac{1}{2}$ cup vegetable oil. In total, the recipe calls for how many cups?

3. Don worked $\frac{5}{8}$ of an hour overtime on Monday, $\frac{3}{8}$ of an hour overtime on Wednesday, and $\frac{7}{8}$ of an hour overtime on Friday. How many hours overtime did Don work for the week?

4. At various times on Tuesday, Tonya sold the following lengths of fabric: 2 feet, 3 feet, 8 feet, and 5 feet. How many yards of fabric did Tonya sell? (One foot equals $\frac{1}{3}$ of a yard.)

5. On a trip to the local health food store, Bill bought $\frac{1}{2}$ pound of sunflower seeds, $\frac{1}{4}$ pound of pumpkin seeds, $\frac{3}{4}$ pound of oat bran, $\frac{3}{4}$ pound of wheat germ, and $\frac{7}{8}$ pound of brown rice. What was the total weight of Bill's purchases?

ANSWERS

1. _____

2. _____

3. _____

4. _____

5. _____

6. Dennis jogs $\frac{3}{4}$ mile every morning and $\frac{1}{3}$ mile every evening. How many miles does Dennis jog each day?

7. Jill spends $\frac{1}{4}$ of her paycheck on food, $\frac{1}{3}$ on rent, $\frac{1}{6}$ on entertainment, and $\frac{1}{5}$ on miscellaneous expenses. What total fraction of her pay does Jill spend?

8. To make a bookcase, Bob cut these pieces from a long board: $\frac{2}{3}$ foot, $\frac{1}{8}$ foot, and $\frac{1}{6}$ foot. How many feet of board were cut?

9. Julian and Charlie bought $\frac{7}{8}$ acre of land on June 1. On June 19, they bought another $\frac{1}{2}$ acre. How much land did they buy altogether?

10. John worked these hours last week:

Monday . . . .$8\frac{1}{2}$ hours

Tuesday . . . .$9\frac{1}{2}$ hours

Wednesday . . . . . .$10\frac{1}{4}$ hours

Thursday . . . .$8\frac{2}{3}$ hours

Friday . . . .$8\frac{1}{2}$ hours

How many total hours did John work during the week?

unit 3.5

Subtracting Fractions and Mixed Numbers

After completing Unit 3.5, you will be able to:

1. Subtract fractions and reduce final answers to their lowest terms.
2. Subtract mixed numbers.

The process of subtracting fractions is very similar to the process of adding fractions. To subtract fractions, follow these steps:

Step 1: Line up the fractions vertically, the smaller fractions under the larger ones.

Step 2: Find the least common denominator (LCD), and raise the fractions to equivalent terms.

Step 4: Subtract the numerators.

Step 5: Reduce the final answer to its lowest terms, if needed.

Example 1

Subtract: $\frac{7}{8} - \frac{3}{4}$.

Solution

We can observe that the LCD is 8. Raise the fractions to 8ths and subtract the numerators:

$$\begin{array}{r} \dfrac{7}{8} = \dfrac{7}{8} \\[2mm] -\dfrac{3}{4} = \dfrac{6}{8} \\[1mm] \hline \dfrac{1}{8} \end{array}$$

$$8 \div 4 = 2;\ 2 \times 3 = 6$$

$$\frac{7}{8} - \frac{6}{8} = \frac{1}{8}$$

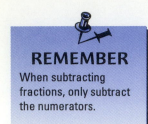
Example 2

Subtract: $\frac{3}{10} - \frac{2}{15}$.

Solution

The LCD is 30. Raise the fractions to 30ths and subtract the numerators:

$$\frac{3}{10} = \frac{9}{30}$$
$$-\frac{2}{15} = \frac{4}{30}$$
$$\frac{5}{30} = \frac{5 \div 5}{30 \div 5} = \frac{1}{6}$$

Divide both terms by 5 to reduce the answer to its lowest terms.

Self-Check

Subtract the following:

(a) $\frac{2}{3} - \frac{1}{3}$ (b) $\frac{7}{8} - \frac{2}{3}$ (c) $\frac{1}{6} - \frac{1}{9}$

Solution

(a) $\frac{1}{3}$ (b) $\frac{5}{24}$ (c) $\frac{1}{18}$

SUBTRACTING MIXED NUMBERS

To subtract mixed numbers, follow these steps:

Step 1: Line up the numbers vertically, the smaller fractions under the larger ones.

Step 2: Find a common denominator and subtract the fractions.

Step 3: Subtract the whole numbers.

Step 4: Reduce the final answer to its lowest terms, if needed.

Example

Subtract $5\frac{3}{4} - 3\frac{1}{3}$.

Solution

The LCD is 12; raise the fractions to equivalent fractions with a common denominator of 12.

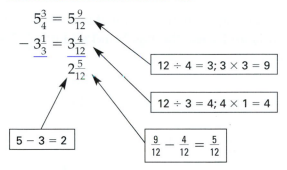

$$5\frac{3}{4} = 5\frac{9}{12}$$
$$-\,3\frac{1}{3} = 3\frac{4}{12}$$
$$2\frac{5}{12}$$

| $12 \div 4 = 3; 3 \times 3 = 9$ |
| $12 \div 3 = 4; 4 \times 1 = 4$ |

| $5 - 3 = 2$ |
| $\frac{9}{12} - \frac{4}{12} = \frac{5}{12}$ |

≈ SUBTRACTING MIXED NUMBERS WHEN BORROWING IS NEEDED

When you subtract mixed numbers, the fraction in the bottom number (the subtrahend) is sometimes larger than the fraction in the top number (the minuend). When this happens, you borrow 1 from the top whole number and add it to the top fraction. Let's look at a couples of examples.

Example 1

Subtract: $8\frac{1}{3} - 5\frac{3}{5}$.

Solution

The LCD is 15; raise the fractions to equivalent fractions with a common denominator of 15.

$$8\frac{1}{3} = 8\frac{5}{15}$$
$$-\,5\frac{3}{5} = 5\frac{9}{15}$$

| $15 \div 3 = 5; 5 \times 1 = 5$ |
| $15 \div 5 = 3; 3 \times 3 = 9$ |

Notice that $\frac{9}{15}$ is larger than $\frac{5}{15}$. Since we cannot subtract $\frac{9}{15}$ from $\frac{5}{15}$, we have to borrow a 1 from the 8; change the 8 to 7 and write the 1 as $\frac{15}{15}$, and add it to the $\frac{5}{15}$: $\frac{15}{15} + \frac{5}{15} = \frac{20}{15}$. Then rewrite the problem to show the borrowing:

$$\overset{7}{\cancel{8}}\frac{5}{15} + \frac{15}{15} = 7\frac{20}{15}$$
$$-\,5\frac{9}{15} \qquad\quad = 5\frac{9}{15}$$
$$2\frac{11}{15}$$

| $7 - 5 = 2$ |

| $\frac{20}{15} - \frac{9}{15} = \frac{11}{15}$ |

Example 2

Subtract: $12\frac{3}{4} - 6\frac{7}{8}$.

Solution
The LCD is 8.

$$12\frac{3}{4} = 12\frac{6}{8} \longleftarrow \boxed{8 \div 4 = 2; 2 \times 3 = 6}$$
$$-6\frac{7}{8} = 6\frac{7}{8}$$

Since $\frac{7}{8}$ is larger than $\frac{6}{8}$, we must borrow a 1 from the 12. Change the 12 to 11 and write the 1 as $\frac{8}{8}$, and add it to $\frac{6}{8}$: $\frac{8}{8} + \frac{6}{8} = \frac{14}{8}$. Rewrite the problem to show the borrowing.

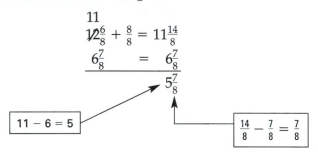

$$\overset{11}{\cancel{12}}\frac{6}{8} + \frac{8}{8} = 11\frac{14}{8}$$
$$6\frac{7}{8} = 6\frac{7}{8}$$
$$5\frac{7}{8}$$

$\boxed{11 - 6 = 5}$ $\boxed{\frac{14}{8} - \frac{7}{8} = \frac{7}{8}}$

≋ SUBTRACTING A MIXED NUMBER FROM A WHOLE NUMBER

To subtract a mixed number from a whole number, follow the steps in the next example.

Example
Subtract: $18 - 12\frac{1}{8}$.

Solution:

Step 1: Write the problem vertically:

$$\begin{array}{r} 18 \\ -12\frac{1}{8} \end{array}$$

Step 2: Borrow a 1 from the whole number and convert the 1 to a fraction equivalent to the fraction in the bottom number; then subtract:

$$\begin{array}{r} 18 = 17\frac{8}{8} \\ -12\frac{1}{8} = 12\frac{1}{8} \\ \hline 5\frac{7}{8} \end{array}$$

REMEMBER

Anytime a number is written over itself, the value is always equal to 1. So, $\frac{12}{12}, \frac{7}{7}, \frac{55}{55}, \frac{500}{500}$, as well as any other number written over itself, are equal to a value of 1.

Self-Check
Subtract the following:

(a) $17\frac{1}{2} - 3\frac{3}{4}$ **(b)** $8\frac{1}{3} - 2\frac{7}{8}$

Solution

(a) $13\frac{3}{4}$ **(b)** $5\frac{11}{24}$

Exercises

Name _____ Date_____

Subtract these fractions. Reduce answers to their lowest terms:

ANSWERS

1. $\dfrac{3}{4}$
   $-\dfrac{1}{2}$

2. $\dfrac{4}{5}$
   $-\dfrac{3}{4}$

3. $\dfrac{7}{8}$
   $-\dfrac{5}{7}$

4. $\dfrac{8}{9}$
   $-\dfrac{2}{3}$

5. $\dfrac{6}{7}$
   $-\dfrac{3}{5}$

6. $\dfrac{4}{7}$
   $-\dfrac{3}{8}$

7. $\dfrac{2}{3}$
   $-\dfrac{1}{6}$

8. $\dfrac{5}{6}$
   $-\dfrac{4}{9}$

9. $\dfrac{3}{8}$
   $-\dfrac{2}{9}$

10. $\dfrac{4}{5}$
    $-\dfrac{5}{8}$

11. $\dfrac{1}{3}$
    $-\dfrac{1}{4}$

12. $\dfrac{8}{9}$
    $-\dfrac{5}{8}$

13. $\dfrac{4}{5}$
    $-\dfrac{2}{7}$

14. $\dfrac{6}{7}$
    $-\dfrac{4}{9}$

15. $\dfrac{5}{6}$
    $-\dfrac{7}{9}$

16. $\dfrac{5}{9}$
    $-\dfrac{3}{8}$

17. $\dfrac{3}{7}$
    $-\dfrac{2}{9}$

18. $\dfrac{1}{2}$
    $-\dfrac{4}{9}$

19. $\dfrac{2}{3}$
    $-\dfrac{7}{26}$

20. $\dfrac{11}{12}$
    $-\dfrac{11}{20}$

1. _____
2. _____
3. _____
4. _____
5. _____
6. _____
7. _____
8. _____
9. _____
10. _____
11. _____
12. _____
13. _____
14. _____
15. _____
16. _____
17. _____
18. _____
19. _____
20. _____

21. $\dfrac{8}{9} - \dfrac{1}{3}$ **22.** $\dfrac{3}{4} - \dfrac{2}{9}$ **23.** $\dfrac{1}{4} - \dfrac{2}{25}$ **24.** $\dfrac{3}{5} - \dfrac{2}{15}$ **25.** $\dfrac{10}{11} - \dfrac{2}{3}$

Subtract these mixed numbers. Reduce answers to their lowest terms:

26. $15\frac{3}{5}$ **27.** $18\frac{5}{6}$ **28.** $12\frac{4}{7}$ **29.** $25\frac{3}{8}$

$\quad\;\; 8\frac{1}{3}$ $\qquad\; 7\frac{3}{4}$ $\qquad 10\frac{1}{2}$ $\qquad 12\frac{2}{3}$

30. $13\frac{7}{8}$ **31.** $34\frac{5}{6}$ **32.** $77\frac{3}{7}$ **33.** $11\frac{2}{5}$

$\quad 12\frac{4}{7}$ $\qquad 31\frac{3}{5}$ $\qquad 45\frac{4}{9}$ $\qquad 10\frac{1}{3}$

34. $24\frac{4}{5}$ **35.** $39\frac{6}{7}$ **36.** $45\frac{1}{8}$ **37.** $15\frac{7}{9}$

$\quad 16\frac{6}{7}$ $\qquad 21\frac{2}{3}$ $\qquad 32\frac{5}{7}$ $\qquad 12\frac{8}{9}$

38. $31\frac{11}{12}$ **39.** 15 **40.** 25

$\quad 20\frac{9}{18}$ $\qquad\; 8\frac{3}{5}$ $\qquad 12\frac{7}{8}$

unit
3.5

Applications

1. Gerri took $\frac{3}{4}$ pound of grapes to work with her. At lunch she ate $\frac{1}{6}$ pound. How much was left?

2. Which is larger, $\frac{7}{8}$ or $\frac{7}{9}$? How much larger?

3. Jeffery ate $\frac{3}{4}$ of a pizza and Clinton ate $\frac{2}{3}$ of a pizza. How much more did Jeffery eat?

4. A cake recipe calls for $\frac{3}{4}$ cup of white sugar and $\frac{1}{8}$ cup of brown sugar. How much more white sugar is needed than brown sugar?

5. When Ellen started her trip, she had a full tank of gas ($\frac{1}{1}$). At the end of the trip, she only had $\frac{1}{8}$ tank. How much gas did she use?

6. Bill McKay owned $10\frac{1}{2}$ acres of land. At his daughter's marriage, Bill gave her $2\frac{1}{4}$ acres as a home site. How much land did Bill have left?

ANSWERS

1. _____

2. _____

3. _____

4. _____

5. _____

6. _____

149

7. Before cooking, a large steak weighed $4\frac{1}{8}$ pounds. After cooking, it weighed $3\frac{4}{5}$ pounds. How much weight was lost in cooking?

7. _____

8. _____

9. _____

10. _____

11. _____

12. _____

8. Lynn bought a bag of flour that weighed 48 ounces. She then used $12\frac{3}{4}$ ounces to bake a cake. How much was left?

9. How many feet would be left if $10\frac{1}{2}$ feet are cut from a board that is $18\frac{1}{8}$ feet long?

10. John bought $12\frac{1}{4}$ acres of land on June 1. He bought another $25\frac{3}{8}$ acres on July 15. He then sold $10\frac{1}{2}$ acres on August 31. How many acres of land did John have left?

11. The stock of ABC Corporation opened on Tuesday morning at $48\frac{1}{2}$ ($48.50) a share. It closed Tuesday evening at $48\frac{1}{8}$. How much value did the stock lose?

12. A piece of material is $35\frac{7}{8}$ yards long. If $12\frac{2}{5}$ yards are sold, how much material is left?

Multiplying Fractions and Mixed Numbers

After completing Unit 3.6, you will be able to:

1. Multiply simple fractions.
2. Multiply mixed numbers.

To multiply fractions, you multiply like terms: numerators times numerators and denominators times denominators. Answers should be reduced to their lowest terms, if needed.

Example 1
Multiply: $\frac{4}{5} \times \frac{3}{4}$.

Solution

Step 1: Multiply like terms of the fractions:

$$\frac{4}{5} \times \frac{3}{4} = \frac{12}{20}$$

Step 2: Reduce to their lowest terms:

$$\frac{12}{20} = \frac{12 \div 4}{20 \div 4} = \frac{3}{5}$$

Example 2

Multiply: $\frac{2}{3} \times \frac{1}{2} \times \frac{3}{4}$.

Solution

Step 1: Multiply like terms:

$$\frac{2}{3} \times \frac{1}{2} \times \frac{3}{4} = \frac{6}{24}$$

Step 2: Reduce to their lowest terms:

$$\frac{6}{24} = \frac{6 \div 6}{24 \div 6} = \frac{1}{4}$$

CANCELLATION

You can simplify the multiplication of many fractions by using **cancellation.** Cancellation can be used when a numerator *and* a denominator can be evenly divided by the same whole number.

Example 1

Multiply using cancellation: $\frac{3}{4} \times \frac{7}{9}$.

Solution

$$\frac{\overset{1}{\cancel{3}}}{4} \times \frac{7}{\underset{3}{\cancel{9}}} = \frac{7}{12}$$

> The 3 and the 9 are both evenly divisible by 3: 3 ÷ 3 = 1; 9 ÷ 3 = 3.

Example 2

Multiply using cancellation: $\frac{4}{5} \times \frac{7}{10}$.

Solution

$$\frac{\overset{2}{\cancel{4}}}{5} \times \frac{7}{\underset{5}{\cancel{10}}} = \frac{14}{25}$$

> The 4 and the 10 are both evenly divisible by 2: 4 ÷ 2 = 2; 10 ÷ 2 = 5.

It should be stressed that cancellation can be used only when a numerator and a denominator are evenly divisible by the same whole number. The numerators and denominators do not have to be in the same fraction, and you can cancel more than once in a problem.

Cancellation is especially helpful when you are multiplying more than two fractions. Let's look at an example.

Example

Multiply $\frac{7}{8} \times \frac{3}{4} \times \frac{16}{21} \times \frac{2}{9}$.

Solution

$$\frac{\overset{1}{\cancel{7}}}{\underset{1}{\cancel{8}}} \times \frac{\overset{1}{\cancel{3}}}{\underset{\underset{1}{2}}{\cancel{4}}} \times \frac{\overset{\overset{1}{2}}{\cancel{16}}}{\underset{\underset{1}{3}}{\cancel{21}}} \times \frac{\overset{1}{\cancel{2}}}{9} = \frac{1}{9}$$

We made quite a few cancellations in this problem. Here is the order in which they were made:

1. 7 and 21: $7 \div 7 = 1$; $21 \div 7 = 3$.
2. 16 and 8: $8 \div 8 = 1$; $16 \div 8 = 2$.
3. 3 and 3: $3 \div 3 = 1$; $3 \div 3 = 1$.
4. 4 and 2: $4 \div 2 = 2$; $2 \div 2 = 1$.
5. 2 and 2: $2 \div 2 = 1$; $2 \div 2 = 1$.

 **HELPFUL HINT**

Sometimes you will have to work with problems such as "what is $\frac{1}{2}$ of $\frac{3}{5}$?" When such a problem is given, the word "of" means to multiply. So, $\frac{1}{2}$ of $\frac{3}{5}$ is

$$\frac{1}{2} \times \frac{3}{5} = \frac{3}{10}$$

REMEMBER

1. When multiplying fractions, multiply only like terms.
2. Use a cancellation only when there is a whole number that can be divided evenly into *both* a numerator and a denominator; *you never cancel using just numerators or denominators.*

Self-Check

Multiply the following. Use cancellation where possible:

(a) $\frac{7}{8} \times \frac{4}{5}$ **(b)** $\frac{8}{9} \times \frac{3}{8}$ **(c)** $\frac{5}{6} \times \frac{1}{5}$ **(d)** $\frac{4}{5} \times \frac{2}{5}$ **(e)** $\frac{6}{7} \times \frac{7}{8}$

Solution

(a) $\frac{7}{10}$ **(b)** $\frac{1}{3}$ **(c)** $\frac{1}{6}$ **(d)** $\frac{8}{25}$ **(e)** $\frac{3}{4}$

 MULTIPLYING A MIXED NUMBER BY A MIXED NUMBER

To multiply a mixed number by another mixed number, follow these steps:

Step 1: Change the mixed numbers to improper fractions.

Step 2: Cancel, where possible, and multiply across.

Step 3: Change answers that are improper fractions to whole numbers or to mixed numbers.

Example

Multiply: $2\frac{1}{3} \times 5\frac{1}{4}$.

Solution

Step 1: Change the mixed numbers to improper fractions:

$$2\frac{1}{3} = \frac{(3 \times 2 = 6) + 1}{3} = \frac{7}{3}$$

$$5\frac{1}{4} = \frac{(4 \times 5 = 20) + 1}{4} = \frac{21}{4}$$

Step 2: Cancel and multiply across:

$$\frac{7}{\overset{3}{\underset{1}{\cancel{3}}}} \times \frac{\overset{7}{\cancel{21}}}{4} = \frac{49}{4}$$

Step 3: Change an improper fraction to a mixed number:

$$\frac{49}{4} = \begin{array}{r} 12 \\ 4\overline{)49} \\ \underline{48} \\ 1 \end{array} = 12\frac{1}{4}$$

Self-Check

Multiply:

$6\frac{2}{3} \times 2\frac{5}{8}$.

Solution

$17\frac{1}{2}$

MULTIPLYING A WHOLE NUMBER BY A MIXED NUMBER

To multiply a whole number by a mixed number, you change the mixed number to an improper fraction and change the whole number to an improper fraction; then you cancel, where possible, and multiply across.

Example 1

Multiply: $5 \times 3\frac{1}{5}$.

Solution

Step 1: Change the whole number and the mixed number to improper fractions:

$$5 \longrightarrow \frac{5}{1}$$

$$3\frac{1}{5} = \frac{(5 \times 3 = 15) + 1}{5} = \frac{16}{5}$$

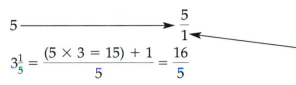

A whole number is changed to an improper fraction by writing the whole number over the number **1**.

Step 2: Cancel and multiply across:

$$\frac{\overset{1}{\cancel{5}}}{1} \times \frac{16}{\underset{1}{\cancel{5}}} = \frac{16}{1} = 16$$

Example 2

Multiply: $24 \times 6\frac{1}{4}$.

Solution

Step 1: Change the whole number and the mixed number to improper fractions:

$$24 \longrightarrow \frac{24}{1}$$

$$6\frac{1}{4} = \frac{(4 \times 6 = 24) + 1}{4} = \frac{25}{4}$$

Step 2: Cancel and multiply across:

$$\frac{\overset{6}{\cancel{24}}}{1} \times \frac{25}{\underset{1}{\cancel{4}}} = \frac{150}{1} = 150$$

Self-Check

Multiply: $48 \times 5\frac{3}{4}$.

Solution

276

unit
3.6

Exercises

Multiply each of these problems. Use cancellation where possible:

ANSWERS

1. $\dfrac{3}{4} \times \dfrac{4}{5}$ 2. $\dfrac{7}{8} \times \dfrac{5}{9}$ 3. $\dfrac{8}{9} \times \dfrac{2}{7}$ 4. $\dfrac{3}{8} \times \dfrac{4}{5}$ 5. $\dfrac{5}{6} \times \dfrac{7}{8}$

6. $\dfrac{1}{5} \times \dfrac{4}{7}$ 7. $\dfrac{3}{5} \times \dfrac{4}{9}$ 8. $\dfrac{2}{3} \times \dfrac{6}{9}$ 9. $\dfrac{1}{4} \times \dfrac{5}{7}$ 10. $\dfrac{2}{7} \times \dfrac{7}{9}$

11. $\dfrac{8}{9} \times \dfrac{3}{4}$ 12. $\dfrac{3}{5} \times \dfrac{5}{9}$ 13. $\dfrac{5}{7} \times \dfrac{7}{8}$ 14. $\dfrac{3}{5} \times \dfrac{7}{9}$ 15. $\dfrac{1}{3} \times \dfrac{4}{5}$

16. $\dfrac{4}{5} \times \dfrac{3}{8}$ 17. $\dfrac{2}{5} \times \dfrac{5}{8}$ 18. $\dfrac{8}{9} \times \dfrac{7}{8}$ 19. $\dfrac{2}{7} \times \dfrac{7}{8}$ 20. $\dfrac{1}{5} \times \dfrac{5}{6}$

1. _____
2. _____
3. _____
4. _____
5. _____
6. _____
7. _____
8. _____
9. _____
10. _____
11. _____
12. _____
13. _____
14. _____
15. _____
16. _____
17. _____
18. _____
19. _____
20. _____

21. $\dfrac{7}{9} \times \dfrac{3}{5}$ **22.** $\dfrac{2}{5} \times \dfrac{1}{2} \times \dfrac{3}{4}$ **23.** $\dfrac{3}{4} \times \dfrac{4}{7} \times \dfrac{7}{9}$ **24.** $\dfrac{4}{5} \times \dfrac{6}{8} \times \dfrac{8}{9}$

25. $\dfrac{4}{7} \times \dfrac{7}{8} \times \dfrac{2}{3}$ **26.** $\dfrac{2}{5} \times \dfrac{5}{8} \times \dfrac{8}{9}$ **27.** $\dfrac{2}{6} \times \dfrac{6}{8} \times \dfrac{3}{4}$ **28.** $\dfrac{3}{7} \times \dfrac{7}{9} \times \dfrac{3}{4}$

26. _____

27. _____

28. _____

29. _____

30. _____

31. _____

32. _____

29. $\dfrac{2}{5} \times \dfrac{4}{7} \times \dfrac{6}{9} \times \dfrac{1}{8}$ **30.** $\dfrac{2}{3} \times \dfrac{5}{6} \times \dfrac{8}{9} \times \dfrac{1}{2}$ **31.** $\dfrac{3}{7} \times \dfrac{7}{9} \times \dfrac{5}{6} \times \dfrac{3}{5}$

33. _____

34. _____

35. _____

36. _____

37. _____

38. _____

39. _____

40. _____

41. _____

42. _____

43. _____

44. _____

45. _____

Multiply each of these problems. Change answers that are improper fractions to whole numbers or to mixed numbers:

32. $5\frac{1}{3} \times 3\frac{1}{2}$ **33.** $7\frac{2}{5} \times 3\frac{1}{8}$ **34.** $6\frac{1}{8} \times 3\frac{2}{7}$

35. $4\frac{2}{5} \times 5\frac{5}{8}$ **36.** $6\frac{2}{3} \times 7\frac{5}{9}$ **37.** $8\frac{6}{7} \times 3\frac{5}{8}$

38. $2\frac{4}{7} \times 4\frac{6}{7}$ **39.** $3\frac{3}{4} \times 2\frac{2}{9}$ **40.** $1\frac{8}{9} \times 5\frac{1}{3}$

41. $6\frac{3}{4} \times 2\frac{7}{8}$ **42.** $2\frac{6}{7} \times 7\frac{4}{5}$ **43.** $4\frac{5}{6} \times 7$

44. $15 \times 6\frac{3}{5}$ **45.** $5 \times 3\frac{3}{4}$

unit 3.6

Name _____ **Date** _____

Applications

1. On Tuesday, the Great Southern Mining Company extracted $\frac{1}{2}$ ton of iron ore. If $\frac{1}{8}$ of this is actually good enough to be used to make iron, how much is usable?

2. What is $\frac{3}{4}$ of $\frac{5}{20}$?

3. Find $\frac{1}{2}$ of $\frac{2}{3}$.

4. Nelson Holmes ran $\frac{1}{2}$ of a $\frac{3}{4}$-mile trail. How far did he run?

5. Which is larger: $\frac{1}{4}$ of $\frac{5}{6}$, or $\frac{1}{5}$ of $\frac{7}{8}$? How much larger?

ANSWERS

1. _____

2. _____

3. _____

4. _____

5. _____

159

6. Joe bought 12 boards, each of which is $5\frac{1}{2}$ feet long. What is the combined length of the 12 boards?

7. Sharon saves $\frac{1}{5}$ of her "take-home pay." How much would she save on a payday when she took home $350?

8. If one Whopper Burger weighs $\frac{3}{8}$ pounds, how much would 12 Whopper Burgers weigh?

9. If carpet sells for $8 a square yard, for how much would $22\frac{1}{2}$ square yards sell?

10. What is the cost of $12\frac{1}{8}$ pounds of ground beef if one pound costs $2?

unit
3.7

Multiplying Large Mixed Numbers

After completing Unit 3.7, you will be able to:

1. Multiply a large mixed number by another large mixed number.
2. Multiply a large whole number by a mixed number.

In Unit 3.6, we multiplied mixed numbers by changing them to improper fractions. Although this method works well for many problems, it can result in improper fractions that are hard to work with if the mixed numbers are large. For example, if you were multiplying $226\frac{3}{8} \times 312\frac{1}{4}$, changing the mixed numbers to improper fractions would give you:

$$\frac{1,811}{8} \times \frac{1,249}{4} =$$

As you can see, these improper fractions have large numerators and are somewhat hard to work with. In such situations, it is usually easier to use the **four-step method.** To multiply mixed numbers using the four-step method, follow these steps:

Step 1: Multiply the two fractions.

Step 2: Multiply each whole number by the opposite fraction.

Step 3: Multiply the whole numbers.

Step 4: Add the results from the first three steps.

Example
Using the four-step method, multiply: $226\frac{3}{8} \times 312\frac{1}{4}$.

Solution

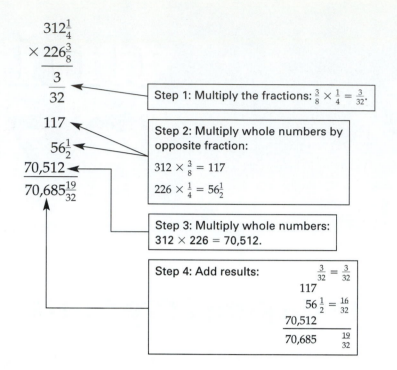

$$312\tfrac{1}{4}$$
$$\times\, 226\tfrac{3}{8}$$

$$\frac{3}{32}$$

Step 1: Multiply the fractions: $\frac{3}{8} \times \frac{1}{4} = \frac{3}{32}$.

$$117$$
$$56\tfrac{1}{2}$$

Step 2: Multiply whole numbers by opposite fraction:

$312 \times \frac{3}{8} = 117$

$226 \times \frac{1}{4} = 56\tfrac{1}{2}$

$$70{,}512$$
$$70{,}685\tfrac{19}{32}$$

Step 3: Multiply whole numbers: $312 \times 226 = 70{,}512$.

Step 4: Add results:

$$\frac{3}{32} = \frac{3}{32}$$
$$117$$
$$56\tfrac{1}{2} = \frac{16}{32}$$
$$70{,}512$$
$$\overline{70{,}685 \quad \frac{19}{32}}$$

≈ MULTIPLYING A LARGE WHOLE NUMBER BY A MIXED NUMBER

To multiply a large whole number by a mixed number, follow the steps in the next example:

Example
Multiply: $160 \times 30\tfrac{3}{4}$.

Solution
Step 1: Multiply the whole number, 160, by the whole number part of the mixed number, 30:

$$\begin{array}{r} 160 \\ \times\ \ 30 \\ \hline 4{,}800 \end{array}$$

Step 2: Multiply the whole number, 160, by the fraction part of the mixed number, $\frac{3}{4}$:

$$\frac{\overset{40}{\cancel{160}}}{1} \times \frac{3}{\underset{1}{\cancel{4}}} = 120$$

Step 3: Add the results from Steps 1 and 2 to get the final product:

$$\begin{array}{r} 4{,}800 \\ +\ \ 120 \\ \hline 4{,}920 \end{array}$$

Answer $= 4{,}920$

Name _____ Date _____

Exercises

Multiply each of these problems. Use cancellation where possible:

ANSWERS

1. $320\frac{1}{2} \times 120\frac{5}{8}$

2. $245\frac{2}{3} \times 270\frac{4}{5}$

3. $420\frac{2}{7} \times 210\frac{5}{6}$

4. $300\frac{7}{8} \times 255\frac{4}{5}$

5. $428\frac{1}{2} \times 218\frac{1}{6}$

6. $485\frac{3}{4} \times 32\frac{4}{5}$

7. $390\frac{1}{8} \times 40\frac{3}{7}$

8. $210\frac{7}{9} \times 810\frac{3}{7}$

9. $120 \times 18\frac{4}{5}$

10. $345 \times 12\frac{1}{8}$

11. $765 \times 25\frac{1}{4}$

12. $530 \times 31\frac{9}{10}$

1. _____
2. _____
3. _____
4. _____
5. _____
6. _____
7. _____
8. _____
9. _____
10. _____
11. _____
12. _____

13. $324 \times 80\frac{1}{6}$

14. $450 \times 16\frac{4}{9}$

15. $386 \times 20\frac{1}{8}$

ANSWERS

13. _____

14. _____

15. _____

16. _____

17. _____

18. _____

19. _____

20. _____

21. _____

22. _____

23. _____

24. _____

25. _____

16. $594 \times 22\frac{5}{6}$

17. $550 \times 25\frac{3}{5}$

18. $300 \times 45\frac{3}{4}$

19. $112 \times 21\frac{3}{16}$

20. $340 \times 18\frac{4}{15}$

21. $262 \times 27\frac{1}{8}$

22. $138 \times 12\frac{5}{6}$

23. $345 \times 21\frac{2}{9}$

24. $565 \times 32\frac{4}{5}$

25. $310 \times 41\frac{6}{7}$

Name _____ Date _____

Applications

1. The cost of a pound of a certain industrial cleaner is $3\frac{5}{8}$. How much would $12\frac{1}{2}$ pounds of the cleaner cost?

2. If carpet costs $12\frac{1}{2}$ a square yard, how much would $25\frac{3}{4}$ square yards cost?

3. If Eddie drives 55 miles per hour, how far would he drive in $10\frac{1}{2}$ hours?

4. If a construction block weighs $8\frac{2}{3}$ pounds, how much would $22\frac{1}{2}$ blocks weigh?

5. Steve earns $8 an hour. How much would he earn in a week in which he worked $39\frac{1}{3}$ hours?

ANSWERS

1. _____

2. _____

3. _____

4. _____

5. _____

6. If a bottle of cola holds $6\frac{1}{2}$ ounces, how many ounces are in 18 bottles?

ANSWERS

6. _____

7. _____

8. _____

9. _____

10. _____

7. The stock of Superior Corporation is selling for $33\frac{1}{3}$ ($33\frac{1}{3}$) per share. How much would 90 shares cost?

8. If it takes $40\frac{1}{8}$ gallons of water to fill a small drum, how many gallons would it take to fill $10\frac{3}{4}$ drums?

9. It takes $3\frac{1}{2}$ yards of fabric to make a certain coat. How many yards of fabric would it take to make 15 coats?

10. What is the cost of $25\frac{1}{2}$ pounds of ground beef if one pound costs $2\frac{1}{4}$?

unit 3.8

Dividing Fractions and Mixed Numbers

After completing Unit 3.8, you will be able to:

1. Divide fractions.
2. Divide a whole number by a fraction.
3. Divide one mixed number by another mixed number.
4. Divide whole numbers by mixed numbers.

When you divide fractions, set up the problem with the dividend first, and the divisor second, like this:

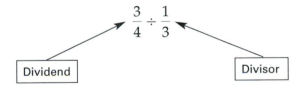

$$\frac{3}{4} \div \frac{1}{3}$$

Dividend   Divisor

To divide fractions, you **invert** (turn upside down) the divisor and then multiply the terms.

Example 1

Divide: $\frac{3}{4} \div \frac{1}{3}$.

Solution

Step 1: Invert the divisor and change the ÷ symbol to a × symbol:

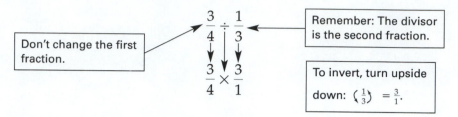

Don't change the first fraction.

Remember: The divisor is the second fraction.

To invert, turn upside down: $\left(\frac{1}{3}\right) = \frac{3}{1}$.

Step 2: Multiply across:

$$\frac{3}{4} \times \frac{3}{1} = \frac{9}{4}$$

Step 3: Change the improper fraction, $\frac{9}{4}$, to a mixed number:

$$\frac{9}{4} = 4\overline{)9} \quad \begin{array}{c} 2 = 2\frac{1}{4} \\ \underline{8} \\ 1 \end{array}$$

Example 2

Divide: $\frac{7}{8} \div \frac{5}{6}$.

Solution

Step 1: Invert the divisor and change the ÷ symbol to a × symbol:

$$\frac{7}{8} \quad \div \quad \left(\frac{5}{6}\right)$$

$$\frac{7}{8} \quad \times \quad \frac{6}{5}$$

Step 2: Cancel and multiply across:

$$\frac{7}{\overset{}{\underset{4}{8}}} \times \frac{\overset{3}{6}}{5} = \frac{21}{20}$$

Step 3: Change the improper fraction, $\frac{21}{20}$, to a mixed number:

$$\frac{21}{20} = 20\overline{)21} \quad \begin{array}{c} 1 = 1\frac{1}{20} \\ \underline{20} \\ 1 \end{array}$$

〰 DIVIDING A WHOLE NUMBER BY A FRACTION

To divide a whole number by a fraction, rewrite the whole number as an improper fraction, invert the divisor, and multiply.

Example 1

Divide: $6 \div \frac{3}{4}$.

Solution

Step 1: Rewrite the whole number as an improper fraction:

$$\frac{6}{1} \div \frac{3}{4}$$

> Remember: To make a whole number an improper fraction, write the whole number over the number 1.

Step 2: Invert the divisor and change the ÷ symbol to a × symbol:

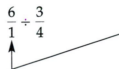

Step 3: Cancel and multiply across:

$$\frac{\cancel{6}^{2}}{1} \times \frac{4}{\cancel{3}_{1}} = \frac{8}{1} = 8$$

Example 2

Divide: $24 \div \frac{7}{8}$.

Solution

Step 1: Rewrite the whole number as an improper fraction:

$$24 = \frac{24}{1}$$

Step 2: Invert the divisor and change the ÷ symbol to a × symbol:

$$\frac{24}{1} \qquad \div \qquad \left(\frac{7}{8}\right)$$

$$\frac{24}{1} \qquad \times \qquad \frac{8}{7}$$

Step 3: Multiply across:

$$\frac{24}{1} \times \frac{8}{7} = \frac{192}{7} \qquad \text{We cannot cancel, so simply multiply across.}$$

Step 4: Change the improper fraction, $\frac{192}{7}$, to a mixed number:

$$\frac{192}{7} = 7\overline{)192} \begin{array}{c} 27 \\ \underline{14} \\ 52 \\ \underline{49} \\ 3 \end{array} = 27\frac{3}{7}$$

Remember:

1. To divide fractions, invert the divisor and multiply.
2. To invert means to turn upside down or to reverse the numerator and the denominator:

$$\frac{2}{3} \text{ inverted } = \frac{3}{2}$$

$$\frac{7}{8} \text{ inverted } = \frac{8}{7}$$

$$\frac{29}{30} \text{ inverted } = \frac{30}{29}$$

Self-Check

Divide:

$18 \div \dfrac{2}{3}$

Solution

27

⁓ DIVIDING MIXED NUMBERS

To divide a mixed number by another mixed number, follow these steps:

Step 1: Change the mixed numbers to improper fractions.

Step 2: Change the ÷ symbol to a × symbol and invert the divisor.

Step 3: Cancel, where possible, and multiply across.

Step 4: Change any answers that are improper fractions to whole numbers or to mixed numbers.

Example 1
Divide: $6\frac{1}{8} \div 4\frac{2}{3}$.

Solution
Step 1: Change the mixed numbers, $6\frac{1}{8}$ and $4\frac{2}{3}$, to improper fractions:

$$6\frac{1}{8} = \frac{(8 \times 6 = 48) + 1}{8} = \frac{49}{8}$$

$$4\frac{2}{3} = \frac{(3 \times 4 = 12) + 2}{2} = \frac{14}{3}$$

Step 2: Change the ÷ symbol to a × symbol and invert the divisor, $\frac{14}{3}$:

Remember: Bring down the first fraction exactly as it is written.	$\dfrac{49}{8}$	÷	$\left(\dfrac{14}{3}\right)$
	$\dfrac{49}{8}$	×	$\dfrac{3}{14}$

Step 3: Cancel and multiply across:

$$\frac{\overset{7}{\cancel{49}}}{8} \times \frac{3}{\underset{2}{\cancel{14}}} = \frac{21}{16}$$

Step 4: Change the improper fraction, $\frac{21}{16}$, to a mixed number:

$$\frac{21}{16} = 16\overline{)21} \quad \begin{array}{r} 1 = 1\frac{5}{16} \\ \underline{16} \\ 5 \end{array}$$

Example 2
Divide: $3\frac{1}{2} \div 5\frac{1}{4}$.

Solution
Change the mixed numbers to improper fractions, invert the divisor, cancel, and multiply:

$$3\frac{1}{2} \div 5\frac{1}{4} = \frac{7}{2} \div \frac{21}{4} = \frac{\overset{1}{\cancel{7}}}{\underset{1}{\cancel{2}}} \times \frac{\overset{2}{\cancel{4}}}{\underset{3}{\cancel{21}}} = \frac{2}{3}$$

≋ DIVIDING WHOLE NUMBERS BY MIXED NUMBERS

To divide a whole number by a mixed number (or a mixed number by a whole number), you change the whole number and the mixed number to improper fractions, invert the divisor, and multiply. Before you multiply, remember to perform any cancellation where possible.

Example 1
Divide: $24 \div 3\frac{1}{3}$.

Solution
Step 1: Change the whole number and the mixed number to improper fractions:

$$24 = \frac{24}{1}$$

$$3\frac{1}{3} = \frac{(3 \times 3 = 9) + 1}{3} = \frac{10}{3}$$

Step 2: Change the $\div$ symbol to a $\times$ symbol and invert the divisor:

$$\frac{24}{1} \qquad \div \qquad \frac{10}{3}$$
$$\downarrow \qquad \downarrow \qquad \downarrow$$
$$\frac{24}{1} \qquad \times \qquad \frac{3}{10}$$

Step 3: Cancel and multiply across:

$$\frac{\overset{12}{\cancel{24}}}{1} \times \frac{3}{\underset{5}{\cancel{10}}} = \frac{36}{5}$$

Step 4: Change the improper fraction, $\frac{36}{5}$, to a mixed number:

$$\frac{36}{5} = 5\overline{)36} = 7\frac{1}{5}$$
$$\underline{35}$$
$$1$$

with quotient 7 above.

Example 2
Divide: $12\frac{4}{5} \div 8$.

Solution
Convert both the mixed number and the whole number to improper fractions:

$$12\frac{4}{5} = \frac{64}{5}$$

$$8 = \frac{8}{1}$$

Invert the divisor, cancel, and multiply:

$$\frac{\overset{8}{\cancel{64}}}{5} \times \frac{1}{\underset{1}{\cancel{8}}} = \frac{8}{5} = 5\overline{)8}\;\begin{matrix}1\end{matrix} = 1\frac{3}{5}$$
$$\frac{5}{3}$$

Self-Check

Divide the following problems:

(a) $12\frac{1}{2} \div 2\frac{2}{9}$

(b) $36 \div 3\frac{1}{3}.$

Solution

(a) $5\frac{5}{8}$

(b) $10\frac{4}{5}$

Exercises

Name _____ Date_____

Divide these problems. Use cancellation where possible:

ANSWERS

1. $\dfrac{3}{4} \div \dfrac{1}{2}$

2. $\dfrac{5}{6} \div \dfrac{2}{3}$

3. $\dfrac{4}{5} \div \dfrac{1}{7}$

4. $\dfrac{1}{8} \div \dfrac{1}{9}$

5. $\dfrac{3}{7} \div \dfrac{9}{10}$

6. $\dfrac{7}{8} \div \dfrac{3}{4}$

7. $\dfrac{1}{3} \div \dfrac{5}{8}$

8. $\dfrac{7}{8} \div \dfrac{5}{8}$

9. $\dfrac{2}{3} \div \dfrac{1}{7}$

10. $\dfrac{1}{5} \div \dfrac{1}{9}$

11. $\dfrac{6}{7} \div \dfrac{2}{5}$

12. $\dfrac{1}{2} \div \dfrac{3}{8}$

13. $\dfrac{7}{8} \div \dfrac{3}{11}$

14. $\dfrac{9}{10} \div \dfrac{2}{15}$

15. $\dfrac{2}{3} \div \dfrac{15}{16}$

1. _____
2. _____
3. _____
4. _____
5. _____
6. _____
7. _____
8. _____
9. _____
10. _____
11. _____
12. _____
13. _____
14. _____
15. _____

16. $\dfrac{21}{23} \div \dfrac{7}{23}$ **17.** $\dfrac{11}{12} \div \dfrac{3}{4}$ **18.** $\dfrac{3}{8} \div \dfrac{5}{12}$

19. $\dfrac{15}{19} \div \dfrac{5}{38}$ **20.** $\dfrac{17}{21} \div \dfrac{17}{21}$ **21.** $\dfrac{1}{2} \div \dfrac{3}{16}$

22. $32\frac{1}{4} \div 5\frac{1}{2}$ **23.** $16\frac{3}{4} \div 10\frac{1}{8}$ **24.** $5\frac{1}{8} \div 2\frac{1}{9}$

25. $8\frac{2}{3} \div 6\frac{1}{4}$ **26.** $10\frac{2}{3} \div 4\frac{1}{6}$ **27.** $6\frac{2}{3} \div 2\frac{2}{5}$

28. $3\frac{3}{4} \div 2\frac{1}{4}$ **29.** $48\frac{1}{2} \div 12\frac{1}{8}$ **30.** $9\frac{3}{5} \div 3\frac{3}{7}$

31. $12 \div 3\frac{1}{3}$ **32.** $24 \div 10\frac{2}{3}$ **33.** $20 \div 4\frac{2}{5}$

34. $6 \div 3\frac{7}{8}$ **35.** $8 \div 3\frac{1}{5}$ **36.** $10 \div 3\frac{1}{3}$

ANSWERS

16. _____
17. _____
18. _____
19. _____
20. _____
21. _____
22. _____
23. _____
24. _____
25. _____
26. _____
27. _____
28. _____
29. _____
30. _____
31. _____
32. _____
33. _____
34. _____
35. _____
36. _____

Applications

1. A carton of fruit juice holds $48\frac{1}{3}$ ounces. How many $6\frac{1}{4}$-ounce glasses can be filled from the carton?

2. If a $10\frac{1}{2}$-ounce box of cereal costs $1.00, how much does one ounce cost? (*Hint:* Change $1.00 to an improper fraction: $\$\frac{1}{1}$.)

3. How many $3\frac{1}{3}$-foot sections of material can be cut from a larger section that is 50 feet long?

4. If it takes $8\frac{1}{2}$ ounces of flour to make a cake, how many cakes can be made from a 68-ounce bag of flour?

5. If Ben drove 255 miles in $4\frac{1}{4}$ hours, how many miles per hour did he average?

6. How many 3-foot pieces of ribbon can be cut from a longer piece measuring $13\frac{1}{2}$ feet?

7. Larry is going to divide $18.50 among his three children. How much is each child's share? (*Hint:* $18.50 = \$18\frac{1}{2}$.)

ANSWERS

1. _____

2. _____

3. _____

4. _____

5. _____

6. _____

7. _____

8. If a $10\frac{1}{4}$-ounce bag of peanuts is divided four ways, what would be the size of each share?

9. Kathy's new car averages $24\frac{1}{2}$ miles to a gallon of gas. On a recent trip, she drove 392 miles. How much gas did she use?

10. After cooking a large steak weighed $18\frac{1}{4}$ ounces. What is the size of one portion if the steak is to serve four people?

8. _____

9. _____

10. _____

section

IV

Decimals: Addition, Subtraction, Multiplication, and Division

unit 4.1

The Meaning of Decimals and Reading Decimals

After completing Unit 4.1, you will be able to:

1. Read decimal numbers.

DECIMALS ARE FRACTIONS

Like a common fraction, a **decimal** stands for a part of a whole unit; this is why decimals are often called **decimal fractions.** But, unlike a common fraction, which can have any number as its denominator, a decimal fraction always has a denominator that is a power of 10. Powers of 10 as you recall are 10, 100, 1,000, 10,000, 100,000, and so on. Also, the denominator of a decimal fraction is not written; it is understood to be 10, 100, 1,000, and so on.

The sign of the decimal is the decimal point. Digits written to the left of the decimal point stand for whole numbers; digits written to the right of the decimal point stand for decimal fractions, or partial units. For example, look at this number:

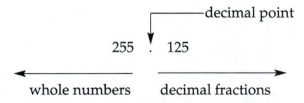

$$\text{255 . 125}$$

whole numbers   decimal fractions

As you remember from Unit 1.1, the value of a digit within a whole number depends on its position, or its "place"; this is called "place value." Place value also applies to decimals or to numbers that are not whole numbers. The following chart shows the place value of whole numbers (from ones to billions) and the place value of decimals (from tenths to billionths).

Billions
Hundred millions
Ten millions
Millions
Hundred thousands
Ten thousands
Thousands
Hundreds
Tens
Ones
• **Decimal point**
Tenths
Hundredths
Thousandths
Ten thousandths
Hundred thousandths
Millionths
Ten millionths
Hundred millionths
Billionths

As you recall, each place to the left of the decimal point is worth ten times the previous place. But, each place to the right of the decimal point is worth one-tenth the place before it. Notice that places to the right of the decimal point end in "ths."

You read a decimal the same way you read a whole number, except that you only read the place value of the right-hand digit.

Example 1

Read .87.

Solution

Read the decimal as you would a whole number: "eighty-seven." Then read the place value of the right-hand digit. The right-hand digit, 7, is in the *hundredths* place; thus, .87 is read as "eight-seven hundredths."

Billions
Hundred millions
Ten millions
Millions
Hundred thousands
Ten thousands
Thousands
Hundreds
Tens
Ones
Decimal point
Tenths
Hundredths
Thousandths
Ten thousandths
Hundred thousandths
Millionths
Ten millionths
Hundred millionths
Billionths

. 8 7

.87 = eighty-seven hundredths

Example 2

Read .675.

Solution

Read the decimal as a whole number: "six hundred seventy-five." Then read the place value of the right-hand digit. The right-hand digit, 5, is in the *thousandths* place; thus, .675 is read as "six hundred seventy-five thousandths."

$.\ 6\ 7\ 5$

.675 = six hundred seventy-five thousandths

Example 3

Read .0002

Solution

The right-hand digit, 2, is in the ten-thousandths place. Thus, the decimal is read as "two ten-thousandths."

$.\ 0\ 0\ 0\ 2$

.0002 = two ten-thousandths

Let's look at how some other decimals should be read and how they would appear as fractions.

Decimal	Read As	Fraction
.7	7 tenths	$\frac{7}{10}$
.16	16 hundredths	$\frac{16}{100}$
.003	3 thousandths	$\frac{3}{1,000}$
.0025	25 ten-thousandths	$\frac{25}{10,000}$
.000175	175 millionths	$\frac{175}{1,000,000}$

≋ MIXED DECIMALS

A **mixed decimal** (mixed number) is a whole number and a decimal fraction—28.4, for example. To read a mixed decimal, first read the whole number, say the word "and" when you reach the decimal point, and then read the decimal.

Example 1

Read 345.28.

Solution

Read the whole number: "three hundred forty-five;" then read the decimal: "and twenty-eight hundredths."

$$345.28 = \text{three hundred forty-five and twenty-eight hundredths}$$

Notice that in reading this mixed decimal, we did not say the word "and" until we reached the decimal point. Remember from Unit 1.1 that the word "and" is not used when you read whole numbers. It is used to separate whole numbers (on the left-hand side of the decimal point) from decimal fractions (on the right-hand side of the decimal point).

Example 2

Read 2,455.5.

Solution

two thousand, four hundred fifty-five and five tenths

Again, notice that "and" was only used to separate the whole number from the decimal.

Work Space

unit
4.1

Exercises

Write these decimals and mixed decimals in words:

1. .8 _____

2. .08 _____

3. .007 _____

4. .15 _____

5. .5125 _____

6. .00008 _____

7. .75431 _____

8. .025 _____

9. .0012 _____

10. .4444 _____

11. .04175 _____

12. .00685 _____

13. .999999 _____

14. 145.8 _____

15. 2.00876 _____

16. 7.85 _____

17. 1.0025 _____

18. 45.0002 _____

19. 767.002 _____

20. 2,345.02 _____

21. 3,458.0521 _____

22. 75,678.1 _____

23. $35.42 _____

24. $5,587.96 _____

25. $34,582.97 _____

Write each of these as a decimal fraction:

26. twenty-five hundredths _____

27. three tenths _____

28. forty-five hundredths _____

29. two hundred fifty-five thousandths _____

30. five hundred thirty-five ten-thousandths _____

31. two and fifty-eight thousandths _____

32. forty-four and nine tenths _____

33. one and forty-four hundredths _____

34. five tenths _____

35. four ten-thousandths _____

36. seven hundred three and four thousandths _____

37. seventy-five and two hundred-thousandths _____

38. four millionths _____

39. ten and eighty-eight ten-thousandths _____

40. seventy-five millionths _____

Write each of these decimals as common fractions:

Decimal	Common Fraction

41. .05 _____

42. .8 _____

43. .45 _____

44. .125 _____

45. .905 _____

46. .0125 _____

47. .25 _____

48. .025 _____

49. .775 _____

50. .0004 _____

ANSWERS

51. _____

52. _____

53. _____

54. _____

55. _____

56. _____

57. _____

58. _____

59. _____

60. _____

Write these common fractions as decimals:

51. $\dfrac{3}{10}$ **52.** $\dfrac{4}{100}$ **53.** $\dfrac{9}{10}$ **54.** $\dfrac{7}{10}$ **55.** $\dfrac{6}{100}$

56. $\dfrac{9}{100}$ **57.** $\dfrac{45}{100}$ **58.** $\dfrac{75}{1,000}$ **59.** $\dfrac{225}{10,000}$ **60.** $\dfrac{19}{100,000}$

Work Space

Applications

Name _____ Date _____

ANSWERS

1. Out of each $100 of operating expenses, Rodriguez Products Company pays $60 in salaries and wages. Write as a decimal fraction the part of operating expenses spent on salaries and wages.

1. _____

2. _____

3. _____

2. Stern Company reported a one-tenth increase in profits over a year ago. Write the increase as a decimal fraction.

4. _____

5. _____

6. _____

3. During March, McGram Company placed 1,000 units of a certain product into production. But, because of spoilage, only 975 units were usable. Write the number of units usable as a decimal fraction of the original amount.

7. _____

4. Champe Summers paid $1,000 down on the purchase of a $10,000 car. Write the down payment as a decimal fraction of the price of the car.

5. Jane McKinney, a sales representative for a candy company, receives a $20 commission on each $100 she sells. Write this as a decimal fraction.

6. Tim McEwen got a raise from $8 an hour to $10 an hour. Write the first wage as a decimal fraction of the new one.

7. A sports coat marked for sale for $100 was reduced by $20. Write the reduction as a decimal fraction.

8. In a recent survey, 7 out of 10 people said that they prefer to live in a warm climate. Write this as a decimal fraction.

ANSWERS

8. _____

9. a. _____

   b. _____

   c. _____

   d. _____

   e. _____

   f. _____

9. Write these as decimal fractions:

   a. 6 out of 10

   b. 8 out of 100

   c. 12 out of 1,000

   d. 300 out of 1,000

   e. 20 out of 10,000

   f. 5,000 out of 10,000

Work Space

SECTION IV   Decimals: Addition, Subtraction, Multiplication, and Division

unit 4.2

Rounding Off Decimals

After completing Unit 4.2, you will be able to:

1. Round decimal numbers to various places.

In Unit 1.3, we rounded off whole numbers to a given place, such as tens, hundreds, thousands, and so on. We learned that rounding off numbers is a way to make the numbers easier both to work with and to remember.

Decimals, such as whole numbers, can be rounded off to make them easier both to work with and to remember. When rounding off decimals, follow the same steps you used to round off whole numbers, except that digits to the right of the place being rounded are dropped (rather than being changed to zeros). Let's see how this is done.

To round a decimal, you:

Step 1: Designate the place to which the decimal is to be rounded: we will call this digit the **rounding place.**

Step 2: Underline the rounding place.

Step 3: Look only at the digit to the immediate right of the rounding place; if this digit is 5 or more, increase the digit in the rounding place by 1; if this digit is 4 or less, do not change the digit in the rounding place.

Step 4: Drop all digits to the right of the rounding place.

Example 1

Round .875 to the nearest tenth.

Solution

Step 1: The place to which the decimal is to be rounded is tenths (remember that the tenths place is one place to the right of the decimal point):

```
                          ┌──────────────┐
                          │ rounding place │
      ┌───────────────────└──────────────┘
      ▼
    .875
```

Step 2: Underline the rounding place:

.<u>8</u>75

Step 3: Look at the digit to the immediate right of the rounding place; it is 7, which is 5 or more. So, the digit in the rounding place is increased by 1:

.$\overset{9}{8}$75 = .9

```
    ┌──────────────────────────┐
    │ The digit to the right of the │
    │ rounding place is 7; increase │
    │ the rounding place by 1:      │
    │ 8 + 1 = 9.                    │
    └──────────────────────────┘
```

Step 4: Drop all digits to the right of the rounding place:

.$\overset{9}{8}\cancel{7}\cancel{5}$ = .9

Example 2

Round .6745 to the nearest hundredth.

Solution

.6<u>7</u>45

```
    ┌────────────────────────────┐
    │ The rounding place is hundredths. │
    └────────────────────────────┘
```

.6<u>7</u>45

```
    ┌──────────────────────────────┐
    │ The digit to the right of the rounding │
    │ place is less than 5; do not change the │
    │ number in the rounding place.          │
    └──────────────────────────────┘
```

.67$\cancel{4}\cancel{5}$ = .67

```
    ┌────────────────────────────┐
    │ Drop all digits to the right of the │
    │ rounding place.                   │
    └────────────────────────────┘
```

Let's look at a few more examples.

25.7881 rounded to the nearest hundredth = 25.79
$325.125 rounded to the nearest cent (hundredth) = $325.13
75.999 rounded to the nearest tenth = 76.0

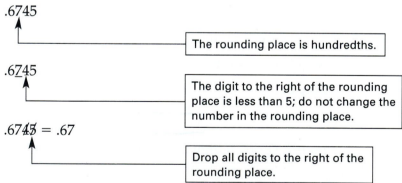

REMEMBER

When rounding decimals, look *only* at the digit to the immediate right of the rounding place; *do not consider any other digit.*

Work Space

Name _____ **Date**_____

Exercises

Round these numbers as shown. The first one is done as an example:

Decimal	Tenths	Hundredths	Thousandths
1. .5984	.6	.60	.598
2. .7775			
3. .8011			
4. .8275			
5. .97347			
6. .65216			
7. 1.25			
8. 4.775			
9. .0685			
10. .222222			
11. .66666666			
12. 17.8962			
13. 275.9951			
14. 2,455.9875			
15. 15,977.9999			

Round each of these to the nearest tenth (1 place):

16. .075

21. 45.666

17. .8875

22. 745.0125

18. .09675

23. 238.9025

19. .6682

24. 2,575.88217

20. .9999

25. 12,455.89165

Round these to the nearest hundredth (2 places):

26. .027

31. 45.6668

27. .9726

32. 76.4582

28. .9012

33. 245.9999

29. .2349

34. 1,259.08212

30. .4567

35. 12,000.345

Round these to the nearest thousandth (3 places):

36. .03877

41. .003872

37. .329767

42. .93276

38. .08254

43. .97254

39. .49875

44. .1287945

40. .098714

45. 12.3457

16. _____
17. _____
18. _____
19. _____
20. _____
21. _____
22. _____
23. _____
24. _____
25. _____
26. _____
27. _____
28. _____
29. _____
30. _____
31. _____
32. _____
33. _____
34. _____
35. _____
36. _____
37. _____
38. _____
39. _____
40. _____
41. _____
42. _____
43. _____
44. _____
45. _____

Round these to the nearest ten-thousandth (4 places):

46. .97563

47. .99999999

48. .29836

49. .0387654

50. .987654

51. .12978678

52. 2.498766

53. 245.69265

54. 98.00345

55. 2,345.98252

Round these to the nearest hundred-thousandth (5 places):

56. .1879768

57. .218096

58. .15678656

59. .1879661

60. .18767867

61. .1781567

62. .21798675

63. .28977891

64. 567.378946

65. 451.789419

Round these to the nearest millionth (6 places):

66. .0078654

67. .7886543

68. .28072687

69. .45870936

70. .21467389

71. 1,238.1875462

72. 5,296.5487611

73. 71.0000876

74. 6,478.4376193

75. 12,345.6715438

46. _____
47. _____
48. _____
49. _____
50. _____
51. _____
52. _____
53. _____
54. _____
55. _____
56. _____
57. _____
58. _____
59. _____
60. _____
61. _____
62. _____
63. _____
64. _____
65. _____
66. _____
67. _____
68. _____
69. _____
70. _____
71. _____
72. _____
73. _____
74. _____
75. _____

Work Space

SECTION IV Decimals: Addition, Subtraction, Multiplication, and Division

Applications

Name _____ Date _____

1. When figuring the total amount to charge a customer, a wholesaler arrived at a figure of $245.6865. Round this figure to even cents (hundredths).

2. A scoop of jelly beans weighs 2.78 pounds. What is the weight of the candy rounded to the nearest tenth?

3. Round 12.4598376 to the nearest (a) ten-thousandth, (b) thousandth, (c) hundredth, (d) tenth.

4. The thickness of a metal rod is .8245 inch. Round the measurement to the nearest hundredth.

5. The value of an ounce of gold is $382.525. Round this figure to even cents.

6. The exact distance from Tom's house to his job is 2.458 miles. Round this figure to the nearest (a) tenth of a mile and (b) hundredth of a mile.

7. The label on a package of ground beef shows that the total weight of the package is 2.75 pounds. Round off the weight to the nearest tenth of a pound.

ANSWERS

1. _____
2. _____
3. a. _____
   b. _____
   c. _____
   d. _____
4. _____
5. _____
6. a. _____
   b. _____
7. _____

8. Following are wholesale prices on items sold by Englewood Company. Round each price to even cents:

Item	Price	
B40	$4.567	_____
D18	$1.019	_____
A10	$0.958	_____
H14	$6.666	_____
A19	$3.458	_____
H10	$0.999	_____

Work Space

unit 4.3

Comparing Decimals

After completing Unit 4.3, you will be able to:

1. Compare decimal numbers to determine their value.

We have learned that the terms of a common fraction can be changed without changing the value of the fraction. For example, we can raise the terms of a common fraction by multiplying both terms by the same nonzero whole number. We can also reduce the terms of a common fraction by dividing both terms by the same number.

The terms of common fractions are changed to make them easier to work with and easier to compare. You can also make decimals easier to work with and easier to compare by changing them so that they will have a common denominator. Decimals have a common denominator when they have the *same number of decimal places*. So, .02 (two hundredths) and .45 (forty-five hundredths) have a common denominator of 100. In the same way, the decimals .005 and .125 have a common denominator of 1,000. Let's look at some examples that compare decimals.

Example 1
Which decimal has a larger value: .385 or .39?

Solution
Step 1: Rewrite the decimals with the same number of places:

$$.385 = .385$$
$$.39 = .390$$

Step 2: Compare the results (the higher number will have the larger value):

The higher decimal is .390, so it has a larger value.

Example 2
Which decimal is smaller: .8 or .08?

Solution

Step 1: Rewrite the decimals with the same number of places:

$$.8 = .80$$
$$.08 = .08$$

Step 2: Look at the results:

$$.80 = \frac{80}{100}$$

$$.08 = \frac{8}{100}$$

.08 is smaller because it is 8 parts out of 100, whereas .80 is 80 parts out of 100.

.08 is smaller than .8.

From these examples we can see that writing zeros to the right of the decimal point does not change the value of the decimal; it simply changes the denominator. As another example, let's take .5. If a zero is added, it becomes .50; if another zero is added, it becomes .500, and so on. These are all ways of writing the same value. We can see this more clearly if we rewrite the decimal as a common fraction and reduce it to its lowest terms:

$$.5 \quad = \frac{5}{10} = \frac{1}{2}$$

$$.50 \quad = \frac{50}{100} = \frac{1}{2}$$

.5, .50, .500 all reduce to $\frac{1}{2}$; thus, they have the same value.

$$.500 = \frac{500}{1,000} = \frac{1}{2}$$

Let's look at another example.

Example

Show that .2 and .20 have the same value.

Solution

$$.2 \quad = \frac{2}{10} = \frac{1}{5}$$

$$.20 = \frac{20}{100} = \frac{1}{5}$$

We can also use models to show that writing zeros after the last digit in a decimal does not change its value. The following models show that .3 has the same value as .30:

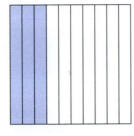

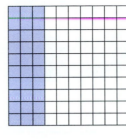

| 3 tenths | = | 30 hundreths |
| .3 | = | .30 |

To add clarity, zeros can also be written to the left of the decimal point. For example, .58 can be written as 0.58. This is a way to show that there are no whole numbers with the decimal. You often see this when dollars and cents are written. To express $.69, for example, you can write $0.69. Writing it this way states definitely that there are no whole dollars, just the 69 cents.

Self-Check

Prove that the following decimals are of equal value:

(a) .9 .90   **(b)** .2 .20 .2000.

Solution

(a) $.9 = \dfrac{9}{10}$

$.90 = \dfrac{90}{100} = \dfrac{9}{10}$

(b) $.2 = \dfrac{2}{10} = \dfrac{1}{5}$

$.20 = \dfrac{20}{100} = \dfrac{1}{5}$

$.2000 = \dfrac{2,000}{10,000} = \dfrac{1}{5}$

Work Space

unit 4.3 Exercises

In each of these problems, circle the decimal with the largest value:

1. .05   .005   .0005   2. .04   .040   .004

3. .70   .07   0.07   4. .8   .808   0.008

5. .55   .055   .050   6. .4   .40   .400

7. .9   .995   .095   8. .02   .20   .020

Prove that the following decimals are of equal value:

9. .6   .60   10. .8   .80   .800

11. .90   .900   0.9   12. .40   .4000

13. .009   0.009   14. .70   0.7   .700

In problems 15–20, underline the decimal with the largest value:

15. .8   .08   .085   16. .65   .665   .065

17. .225   .3   .68   18. .002   .200   0.002

19. .35   .305   0.305   20. .6   .06   .60

In problems 21–24, underline the decimal with the smallest value:

21. .02   .002   .0002   22. .004   .04   0.0004

23. .9   .99   0.99   24. .995   .990   .900

Work Space

SECTION IV   Decimals: Addition, Subtraction, Multiplication, and Division

Applications

Name _____ Date _____

1. On Tuesday, Reggi jogged .8 miles and Colin jogged .75 miles. Who jogged farther? How much farther?

2. Which is larger, eight-tenths of a dollar or three-fourths of a dollar?

3. Part of Dominick's science project was to compare the weights of two test tubes of a heavy liquid metal. One weighed .84 pound and the other weighed .9 pound. Which one weighed more?

4. Harold's height is 5.8 feet, whereas Sam's height is 5.64 feet. Who is taller?

5. Which is larger, .095 or .0925?

6. A bag of brand A sunflower seeds holds 12.2 ounces, whereas a bag of brand B sunflower seeds holds 12.25 ounces. If the cost of each is the same, which is the better value?

7. Which is smaller, .020 or .022?

8. Prove that .8 and .800 have the same value.

ANSWERS

1. _____
2. _____
3. _____
4. _____
5. _____
6. _____
7. _____

Work Space

SECTION IV Decimals: Addition, Subtraction, Multiplication, and Division

unit 4.4

Adding and Subtracting Decimals and Mixed Numbers

After completing Unit 4.4, you will be able to:

1. Add decimals and mixed decimals.
2. Subtract decimals and mixed decimals.

 ## ADDING DECIMALS AND MIXED NUMBERS

RULE

To add decimals, line up the decimal points in a vertical line, then add, following the rules for addition of whole numbers.

Adding decimals is very similar to adding whole numbers. When adding decimals, however, you must line up the decimal points in a straight line. By lining up the decimal points, you automatically line up the place values of the problem. Use the rule to the left to help you when adding decimals and mixed decimals:

Example 1
Add: .521 + .455.

Solution

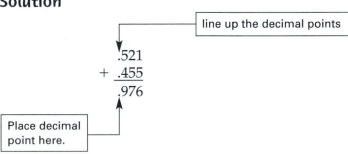

line up the decimal points

```
   .521
+  .455
   .976
```

Place decimal point here.

Example 2
Add: 225.58 + 14.25.

Solution

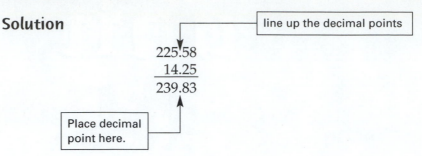

225.58
 14.25
239.83

If you are adding decimals that do not have the same number of places, you can make the problem easier to work with by filling in empty spaces with zeros (adding zeros to the right of the decimal point does not change the value of the number).

Example 1
Add: .02 + .564 + .233.

Solution

.02   .020 ← | Fill in empty space with a zero.
.564 = .564
.233   .233
           .817

Example 2
Add: 2.99 + .7 + .821.

Solution

| Fill in empty spaces with zeros. |

2.99   2.990
 .7 =   .700
 .821   .821
          4.511

Self-Check
Add the following:

(a) .02 + .55 + .7785   **(b)** 15.5 + 125.585 + 5.

Solution
(a) 1.3485   **(b)** 146.085

~~~~ SUBTRACTING DECIMALS AND MIXED DECIMALS

To subtract decimals and mixed decimals, set up the problem as you did for addition; that is, line up the decimal points in a vertical line and subtract as you would with whole numbers.

Example 1
Subtract: 48.75 − 22.61.

Solution

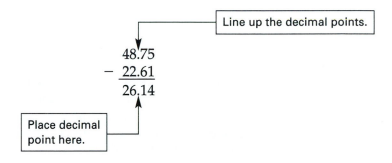

Check: Remember to check a subtraction problem, add the difference (the answer) to the subtrahend (the second number); the sum will equal the minuend (the top number).

$$\begin{array}{r} 26.14 \\ + \ 22.61 \\ \hline 48.75 \end{array}$$

Example 2
Subtract: 345.52 − 31.5.

Solution
Line up the decimal points correctly:

$$\begin{array}{r} 345.52 \\ - \quad 31.5 \\ \hline \end{array}$$

Fill in the empty space with a zero and bring the decimal point into the answer:

$$\begin{array}{r} 345.52 \\ - \quad 31.50 \end{array} \quad \longleftarrow \boxed{\text{Fill in the empty space.}}$$

Subtract as with whole numbers:

$$\begin{array}{r} 345.52 \\ - \quad 31.50 \\ \hline 314.02 \end{array}$$

Check:

$$\begin{array}{r} 314.02 \\ + \quad 31.50 \\ \hline 345.52 \end{array}$$

Example 3
Subtract: 78.51 − 12.39.

Solution

$$
\begin{array}{r}
78.\overset{4\ \ 1}{\cancel{5}1} \\
-\ 12.\ 39 \\
\hline
66.12
\end{array}
$$

Remember to borrow because you cannot subtract 9 from 1.

Check:

$$
\begin{array}{r}
66.12 \\
+\ 12.39 \\
\hline
78.51
\end{array}
$$

We have stressed that zeros can be added to the right of the decimal point without changing the value of the decimal number. If in a subtraction problem the minuend (the top number) has fewer places than the subtrahend (the bottom number), you *must* add zeros to allow you to subtract. For example, how would you subtract 4.2 − .375?

First, set up the problem with the decimal points lined up:

$$
\begin{array}{r}
4.2 \\
-\ \ .375
\end{array}
$$

Then you would fill in the empty spaces with zeros, borrow, and subtract.

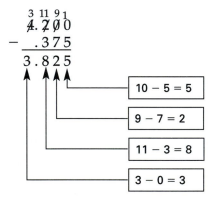

$$
\begin{array}{r}
\overset{3\ \ 11\ \ 9\ \ 1}{4.\cancel{2}\cancel{0}0} \\
-\ \ .375 \\
\hline
3.825
\end{array}
$$

10 − 5 = 5

9 − 7 = 2

11 − 3 = 8

3 − 0 = 3

Let's look at another example.

Example
Subtract: 5.25 − .0175.

Solution
Set up the problem correctly:

$$
\begin{array}{r}
5.25 \\
-\ \ .0175
\end{array}
$$

Fill in the empty spaces with zeros and subtract:

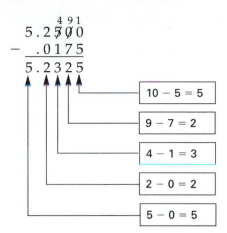

$$5.2\overset{4}{5}\overset{9}{0}\overset{1}{0}$$
$$-\ .0175$$
$$5.2325$$

10 − 5 = 5
9 − 7 = 2
4 − 1 = 3
2 − 0 = 2
5 − 0 = 5

Check:

$$5.2325$$
$$+\ .0175$$
$$5.2500$$

REMEMBER

When adding and subtracting decimal numbers, the numbers are arranged in columns with the decimal points directly under one another.

Self-Check

Subtract:

6.175 − .0245.

Solution

6.1505

Work Space

Exercises

unit 4.4

Add each of these sets of numbers:

1. .2 + .23 + .456

2. .24 + .09 + .34

3. .34 + .381 + .015

4. .00125 + .5812

5. .237 + .3867 + .128

6. .37 + .276 + .8756

7. .8749 + .2 + .47

8. .9215 + .04 + .0004

9. .2458 + .38765 + .9

10. .125 + .78 + .025

11. 2.5 + .004 + .9854

12. 12.58 + 4.589 + 2.983

ANSWERS

1. _____
2. _____
3. _____
4. _____
5. _____
6. _____
7. _____
8. _____
9. _____
10. _____
11. _____
12. _____

13. .2 + 1.2 + 12.875

14. 1.99 + 2.3 + 4.7

13. _____
14. _____
15. _____
16. _____
17. _____
18. _____
19. _____
20. _____
21. _____
22. _____
23. _____
24. _____
25. _____

15. 5.4 + .004 + .0009

16. 12.5 + 25.09 + 67.801

17. 125.8 + 17.51 + 2.0025

18. .00002 + 15 + 2.5

19. 33.89 + 125.01 + .8 + 1.25

20. 65.8 + 21.89 + 234.7

21. $45.91 + $.59

22. $1,234.56 + $34.50

23. $.69 + $2.34 + $235.98

24. $564.89 + $1,245.89

25. $75, 617.84 + $21,319.45 + $12.58 + $125.42

Subtract each of the following problems:

26. 1.2 − .0125　　　**27.** 3.23 − .04　　　**28.** .88 − .09

29. 2.75 − 1.65　　　**30.** 12.58 − 4.75　　　**31.** 35.6 − 10.25

32. 4.675 − 3.789　　　**33.** 45.328 − 21.145　　　**34.** 8 − 2.5

35. .875 − .00125　　　**36.** 32 − .0092　　　**37.** 4.88 − .995

38. 125.87 − 12.365　　　**39.** 458.45 − 4.675　　　**40.** 4.59 − .075

41. 15 − 12.45　　　**42.** 30 − 5.1025　　　**43.** 21.5 − 10.475

44. 55 − 12.65　　　**45.** 225 − 21.62　　　**46.** 65 − 45.75

26. _____
27. _____
28. _____
29. _____
30. _____
31. _____
32. _____
33. _____
34. _____
35. _____
36. _____
37. _____
38. _____
39. _____
40. _____
41. _____
42. _____
43. _____
44. _____
45. _____
46. _____

Subtract the following problems:

47. 2.8 from 3.45

48. 4.785 from 8.9

49. 10.5 from 12.25

50. 18.03 from 42.005

51. 23.0075 from 55.8

52. 8.995 from 12

53. 45.87 from 778.276

54. 2.5 from 15

55. 17.00005 from 50

56. 2.675 from 6.5

57. $25.75 from $100

58. $.69 from $5

59. $21.58 from $65

60. $2.38 from $10

47. _____
48. _____
49. _____
50. _____
51. _____
52. _____
53. _____
54. _____
55. _____
56. _____
57. _____
58. _____
59. _____
60. _____

Applications

Name _____ Date_____

1. The Simmons Company paid these bills for the week ending March 12:

   Rent $600.00
   Utilities 921.43
   Salaries 678.91
   Miscellaneous 97.59

   What was the total amount paid for expenses?

2. On March 12, the bank account of the Hume Fitzsimmons Company had a balance of $4,567.87. On the same date, the company's bookkeeper made a deposit of $3,546.18 and wrote two checks, one for $456.98 and the other for $346.98. What was the balance of the account after these transactions?

3. Joyce jogged 6.5 miles on Tuesday, 6.8 miles on Wednesday, and 4.75 miles on Friday. How many miles did she jog during the week?

4. Bill's salary for the week was $345.85. But, his employer deducted these amounts:

   Federal income taxes $38.10
   State income taxes 12.45
   Social security taxes 24.48
   Health insurance 14.67

   What was Bill's salary after all deductions were taken out?

ANSWERS

1. _____
2. _____
3. _____
4. _____

5. Technical Products, Inc. had sales of $1,234,568.89 for the second quarter of this year. Expenses of $832,456.35 were reported for the same period. What was the amount of profit earned by the company? (*Hint*: Sales − expenses = net profit.)

6. The Fabric Store bought 125 yards of a certain fabric. On Saturday, the following separate sales of the fabric were made: 10.5 yards, 25.8 yards, and 5.4 yards. (a) How many yards of the fabric were sold and (b) how many yards of the fabric remain in stock?

7. During a special sale, the Jones family bought a stereo for $499.95 and a TV set for $699.99. They made a down payment of $150 and agreed to pay the balance in installments. (a) What was the total amount of the purchase and (b) what amount is owed after the down payment?

8. Gregg Blair bought the following books for this semester:

   History $39.95
   Business math 12.45
   Science 24.89
   English 29.99

   If Gregg had $125 with him to buy the books, how much did he have left after he bought them?

9. At 8 o'clock, Kathy and Bob set out on a hiking trail that was 29.5 miles long. After two hours, the couple had traveled 6.8 miles. How much farther did they have to go?

10. On Tuesday, Sam Davis sold 50 shares of MIT stock for $1,200. What was his profit on the sale if he had bought the stock earlier for $879.97?

unit 4.5

Multiplying Decimals

After completing Unit 4.5, you will be able to:

1. Multiply decimal numbers and point off the correct number of places in the product.

There are many real-world business applications that call for the multiplication of decimal numbers. For example, if you want to buy 3 pounds of ground beef, what would be the total cost if one pound costs $1.99? In the same way, what would be your total earnings if you worked 40 hours last week and your hourly wage is $6.50? These situations, along with countless thousands of others each business day, call for decimal numbers to be multiplied.

To multiply numbers that have decimal points, follow these steps:

Step 1: Set up the problem and multiply the numbers as if they were whole numbers; that is, ignore the decimal points when you are multiplying.

Step 2: Count the number of decimal places to the right of the decimal point in the multiplicand *and* the multiplier.

Step 3: Starting from the right-hand digit in the product, move the decimal point to the left the number of places you counted in Step 2, and place the decimal point in this position; this is called **pointing off places.**

Example 1
Multiply: 12.25 × 6.5.

Solution
Set up the problem and multiply as if the numbers were whole numbers:

$$
\begin{array}{r}
12.25 \\
\times \ \ \ 6.5 \\
\hline
6125 \\
7350 \ \ \\
\hline
79625
\end{array}
$$

Count the number of decimal places to the right of the decimal point in the multiplicand and the multiplier:

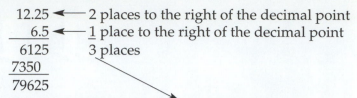

12.25 ← 2 places to the right of the decimal point
6.5 ← 1 place to the right of the decimal point
6125
7350
79625
3 places

Since the multiplicand and the multiplier have 3 places to the right of the decimal point, the product must have 3 places to the right of the decimal point. Move the decimal point 3 places to the left in the product (point off 3 places):

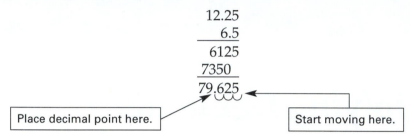

12.25
6.5
6125
7350
79.625

Place decimal point here.   Start moving here.

Example 2
Multiply: 24.25 × 25.

Solution

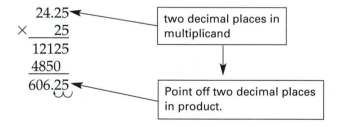

24.25 — two decimal places in multiplicand
× 25
12125
4850
606.25 — Point off two decimal places in product.

Answer = 606.25

When multiplying some decimal numbers, you will have to use zeros in the product to give you the appropriate number of decimal places.

Example 1
Multiply: .28 × .04.

Solution

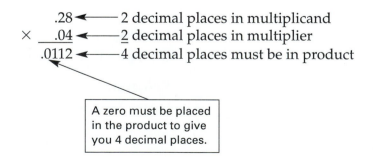

.28 ← 2 decimal places in multiplicand
× .04 ← 2 decimal places in multiplier
.0112 ← 4 decimal places must be in product

A zero must be placed in the product to give you 4 decimal places.

Answer = .0112

Example 2
Multiply: $6.08 \times .0005$.

Solution

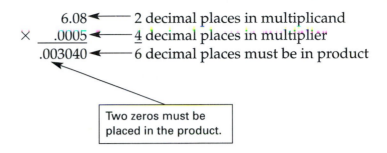

$$
\begin{array}{r}
6.08 \quad \longleftarrow \text{2 decimal places in multiplicand} \\
\times \quad \underline{.0005} \quad \longleftarrow \text{4 decimal places in multiplier} \\
.003040 \quad \longleftarrow \text{6 decimal places must be in product}
\end{array}
$$

Two zeros must be placed in the product.

Answer = .003040

REMEMBER

When multiplying decimal numbers, do the multiplication first; then point off as many decimal places in the product as there are places in both the multiplicand and the multiplier combined.

Self-Check

Multiply the following:

(a) $295 \times .07$ **(b)** $455 \times .008$ **(c)** $.145 \times .12$.

Solution

(a) 20.65   **(b)** 3.64   **(c)** .0174

Work Space

Name _____ Date _____

Exercises

Multiply each of these problems:

1. 225 .09	**2.** 375 .25	**3.** 90 .05	**4.** 850 2.5

5. 127 .2	**6.** 18.5 12	**7.** 176 .09	**8.** 436 5.8

9. 593 .42	**10.** 2,487 3.5	**11.** 2,398 .88	**12.** 4,597 .075

13. 25.38 12	**14.** 45.87 45	**15.** 125.6 3.4	**16.** 30.05 .09

17. 236.87 .14	**18.** 125.854 2.41	**19.** 345.9 .075	**20.** 25.9 .03

21. .125 .008	**22.** .458 .11	**23.** .975 .072	**24.** .0125 .08

ANSWERS

1. _____
2. _____
3. _____
4. _____
5. _____
6. _____
7. _____
8. _____
9. _____
10. _____
11. _____
12. _____
13. _____
14. _____
15. _____
16. _____
17. _____
18. _____
19. _____
20. _____
21. _____
22. _____
23. _____
24. _____

25. 34.58
    6.5

26. 56.4
    .075

27. 2.87
    .75

28. .458
    .28

29. .185
    .22

30. 265.9
    .601

31. 5.7632
    3.45

32. 245.8
    1.8

33. 345.07
    .125

34. 498.12
    3.78

35. 456.18
    .0009

36. 873.006
    2.05

37. 345.79
    .125

38. 984.91
    3.75

39. 2,736.5
    .08

40. 3,456.7
    12

41. $24.75
    15

42. $35.89
    75

43. $87.12
    90

44. $568.56
    50

45. $567.38
    125

46. $37.125
    980

47. $91.475
    525

Answers list:

25. _____
26. _____
27. _____
28. _____
29. _____
30. _____
31. _____
32. _____
33. _____
34. _____
35. _____
36. _____
37. _____
38. _____
39. _____
40. _____
41. _____
42. _____
43. _____
44. _____
45. _____
46. _____
47. _____
48. _____
49. _____
50. _____
51. _____
52. _____
53. _____
54. _____
55. _____

48. $3,800 \times \$.79$

49. $4,582 \times \$.58$

50. $500 \times \$1.95$

51. $8,385 \times \$.09$

52. $10.5 \times \$1.25$

53. $8.25 \times \$.29$

54. $5,118 \times \$.30$

55. $12.75 \times \$1.49$

Name _____ **Date**_____

Applications

1. Sharon House bought 4.5 pounds of chopped meat at a cost of $1.29 a pound. How much did she pay for the meat?

2. Pylant Products, Inc. sold 4,000 pounds of chemical lawn food at $.99 a pound. What was the total amount of the sale?

3. David Valdez earns $5.78 an hour. Last week he worked a total of 39 hours. What was his salary for the week?

4. What is the total cost of five cans of tuna fish, if one can costs $1.12?

5. A local rental car agency advertises that their rates are $25 a day plus $.10 a mile. What would be the total cost of a two-day rental in which a total of 421 miles was driven?

6. The Center Company is going to recarpet their main office. The office is 556.5 square yards and the cost of the carpet is $12.95 a square yard. What will be the total cost of the carpet?

7. If one canister of baking powder weighs 246.8 grams, what would five canisters weigh?

ANSWERS

1. _____

2. _____

3. _____

4. _____

5. _____

6. _____

7. _____

8. Tom is considering buying a VCR that is advertised for $50 down with the balance payable in 12 monthly installments of $30 each. What would be the total cost of the VCR?

9. Katherine Kaye works as a salesperson for Swift Food Products Company. She receives a commission of $.05 per pound on all pork products she sells. Last week she sold 2,500 pounds of pork. What was her commission on this product?

10. On a day when Coca-Cola stock was trading for $78\frac{1}{4}$ ($78.25 per share), Sam McNiel bought 50 shares. What was the total amount of his purchase?

Work Space

228

Multiplying Decimals Using Shortcuts

After completing Unit 4.6, you will be able to:

1. Shortcut multiply decimals by a power of 10.
2. Shortcut multiply decimal numbers by 25 and 50.

   In earlier units, we learned that there are certain shortcuts that can save you time and increase your accuracy when multiplying whole numbers. There are also certain shortcuts that can save you time and increase your accuracy when multiplying decimal numbers. In this unit, we are going to look at some common shortcuts to multiplying decimal numbers.

〰 MULTIPLYING DECIMAL NUMBERS BY A POWER OF 10

Remember that the powers of 10 (or multiples of 10) are 10, 100, 1,000, 10,000, 100,000, and so on. To multiply a number that has a decimal point by a power of 10, follow these two steps:

Step 1: Count the number of zeros in the multiplier (the power of 10 will be the multiplier).

Step 2: Move the decimal point in the multiplicand the same number of places to the right as there are zeros in the multiplier.

Example 1
Multiply: 78.25×100.

Solution

$$78.25 \times 100 = 78.25 = \mathbf{7,825}$$

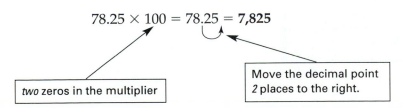

two zeros in the multiplier

Move the decimal point *2* places to the right.

Example 2

Multiply: $28.4 \times 1,000$.

Solution

$$28.4 \times 1,000 = 28.400 = \mathbf{28,400}$$

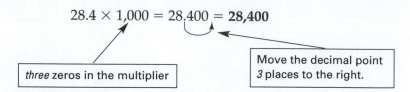

three zeros in the multiplier

Move the decimal point
3 places to the right.

Notice in Example 2 that we had to add two zeros to the product as place-holders to allow us to move the decimal point the correct number of places to the right. Let's look at a few additional examples.

$$356.71 \times 10 = 356.71 = \mathbf{3,567.1}$$

(move 1 place to the right)

$$2.4 \times 100 = 2.40 = \mathbf{240}$$

(move 2 places to the right)

$$245.88 \times 10,000 = 245.8800 = \mathbf{2,458,800}$$

(move 4 places to the right)

Self-Check

Multiply these numbers:

(a) 89.6×10 **(b)** 345.455×100.

Solution

(a) 896   **(b)** 34,545.5

MULTIPLYING BY 25 AND 50

To multiply a number by 25, first multiply by 100 (move the decimal point 2 places to the right), then divide by 4. For example, to multiply 24.48 by 25:

$$24.48 \times 100 = 24.48$$

$$2,448 \div 4 = \mathbf{612}$$

To multiply by 50, first multiply by 100 and then divide by 2. For example, to multiply 144.8 by 50:

$$144.8 \times 100 = 144.80$$

$$14,480 \div 2 = \mathbf{7,240}$$

REMEMBER

To multiply a decimal number by a power of 10, move the decimal point in the multiplicand the same number of places to the right as there are zeros in the multiplier.

Work Space

Exercises

Name _____ Date _____

Multiply each of these problems in your head:

1. 24.5 × 10

2. 45.87 × 100

3. .88 × 10

4. 25.05 × 10

5. 55.8 × 100

6. .8 × 1,000

7. .025 × 100

8. 1.2 × 1,000

9. .67 × 10

10. .002 × 1,000

11. 245.6 × 100

12. 24.09 × 10

13. 1,255.8 × 10

14. $.02 × 1,000

15. $1.55 × 100

16. $.03 × 10,000

17. $125.00 × 10

18. $2.05 × 1,000

19. $145.56 × 100

20. $.125 × 1,000

21. $1.25 × 1,000,000

22. $.375 × 100

ANSWERS

1. _____
2. _____
3. _____
4. _____
5. _____
6. _____
7. _____
8. _____
9. _____
10. _____
11. _____
12. _____
13. _____
14. _____
15. _____
16. _____
17. _____
18. _____
19. _____
20. _____
21. _____
22. _____

Using shortcuts, multiply each of these problems by 25:

23. 448.48

24. 346.16

25. 2,516.32

26. $1,648.64

27. $1,216.36

28. $440.40

Using shortcuts, multiply each of these problems by 50:

29. 252.10

30. 504.62

31. 290.64

32. $.44

33. $1.48

34. $2,464.22

Using shortcuts, multiply these problems. Do them in your head where possible:

Multiplicand	Multiplier	Product
35. 224.8	25	
36. 464.2	50	
37. 2,488.26	50	
38. 6,822.28	25	
39. 640.2	25	
40. 677.8	10	
41. 2.45	1,000	
42. 6.789	100	
43. 36.8	10,000	
44. $345.68	100	
45. $.08	100,000	
46. $2.45	10,000	
47. $1.375	100,000	
48. $.125	1,000,000	
49. $4,569.02	1,000	
50. $331.09	10,000	

Applications

Name _____ **Date** _____

1. The Lennox Products Company bought 10,000 pounds of a certain raw material. If the cost per pound was $.18, what was the total cost of the raw material?

2. Find the price of 100 pounds of ground beef if one pound costs $1.99.

3. The River Road School bought 1,000 pencils at a cost of $.08 each. How much did all the pencils cost?

4. Robert Dodd gets a $.03 sales commission for each pound of meat he sells. What would be his commission in a month when he sells 10,000 pounds of this product?

5. Charter Bookkeeping Service bought ten word processors at a cost of $948.50 each. What was the total amount paid for the word processors?

6. A train averaging 124.8 miles per hour will travel how far in 25 hours?

7. If Dave Alvarez earns $8.55 an hour, how much will he make in 10 hours?

ANSWERS

1. _____

2. _____

3. _____

4. _____

5. _____

6. _____

7. _____

8. What is the cost of 50 cases of cola if one case costs $2.24?

ANSWERS

8. _____

9. _____

10. _____

9. Susan Lane is a sales representative for a major book publisher. Susan is paid $.255 for each mile she drives her personal car. Last year she drove a total of 10,000 miles on company business. How much was she paid for mileage?

10. If a bottle of vitamin E costs $6.95, what is the cost of 1,000 bottles?

unit 4.7

Dividing Decimals

After completing Unit 4.7, you will be able to:

1. Divide a decimal number by a whole number.
2. Divide a decimal by a decimal.

You divide decimal numbers the same way you divide whole numbers. However, you must determine where to place the decimal point in the quotient. (Remember that the quotient is the answer in a division problem.) Where you place the decimal point in the quotient depends on whether your divisor is a whole number or a decimal. Let's look at both situations.

THE WHOLE NUMBER DIVISOR

When the divisor is a whole number, and the dividend has a decimal point, you place the decimal point in the quotient directly above the decimal point in the dividend. Then divide as you would with whole numbers.

Example
Divide: 71.68 by 8.

Solution
Move the decimal point into the quotient:

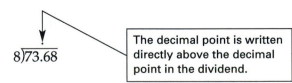

$8\overline{)73.68}$

The decimal point is written directly above the decimal point in the dividend.

Divide as usual:

$$
\begin{array}{r}
9.21 \\
8\overline{)73.68} \\
\underline{72} \\
1\ 6 \\
\underline{1\ 6} \\
8 \\
\underline{8} \\
0
\end{array}
$$

Notice in this example that the numbers divided evenly; that is, there was no remainder. When the numbers do not divide evenly (there is a remainder), there are two ways to handle the remainder:

1. State the remainder as a fraction, as we did in Unit 2.3 when we were dividing whole numbers; or
2. Identify a position to which the decimal is to be rounded; add enough zeros to the dividend to allow you to carry the division to one place beyond the rounding place; then round off. This is the preferred procedure in business. Let's see how it works.

Example

Divide: 76.8 by 18 and round the quotient correctly to the nearest hundredth. (Remember that hundredths is 2 decimal places to the right of the decimal point.)

Solution

Set up the problem and move the decimal point into the quotient:

$$
18\overline{)76.8}
$$

Add enough zeros to the dividend to allow you to carry the division to one place beyond the rounding place. Since you are rounding correctly to hundredths (2 places), you must have 3 decimal places to the right of the decimal point:

$$
\begin{array}{r}
4.266 \\
18\overline{)76.800} \\
\underline{72} \\
4\ 8 \\
\underline{3\ 6} \\
1\ 20 \\
\underline{1\ 08} \\
12
\end{array}
$$

Add two zeros to carry out the division to 3 places.

Now round the quotient correctly to hundredths:

$$4.266 = 4.27$$

≋ THE DECIMAL DIVISOR

When your divisor has a decimal point, follow these steps:

Step 1: Move the decimal point in the divisor as many places to the right as needed to make it a whole number.

Step 2: Move the decimal point in the dividend the **same number** of places as you did in the divisor, attaching zeros if necessary.

Step 3: Place the decimal point in the quotient directly above the newly located decimal point in the dividend.

Step 4: Divide.

Example 1

Divide: 1.44 by .12.

Solution

$$.12.\overline{)1.44.}$$

Move the decimal point 2 places to the right in the divisor and 2 places to the right in the dividend.

Move the newly located decimal point into the quotient, and divide:

```
        12.
   12)1 44.
      1 2
        24
        24
         0
```

The decimal point is placed above its newly located place in the dividend.

Example 2

Divide 135 by 5.7 and round the quotient correctly to the nearest tenth.

Solution

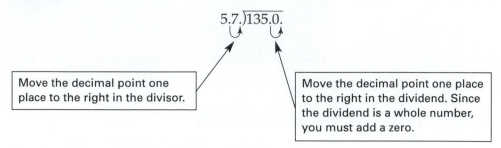

$5.7.\overline{)135.0.}$

Move the decimal point one place to the right in the divisor.

Move the decimal point one place to the right in the dividend. Since the dividend is a whole number, you must add a zero.

Now divide as usual. (Remember that we are rounding the quotient to the nearest tenth.):

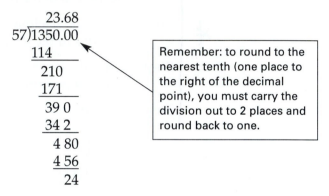

```
        23.68
   57)1350.00
       114
       210
       171
        39 0
        34 2
         4 80
         4 56
           24
```

Remember: to round to the nearest tenth (one place to the right of the decimal point), you must carry the division out to 2 places and round back to one.

Answer: $23.68 = 23.7$

Self-Check

Divide 12.5 by 4.2 and round to the nearest tenth.

Solution

3.0

Example 3

Divide 675 by 3.28 and round the quotient to the nearest thousandth.

Solution

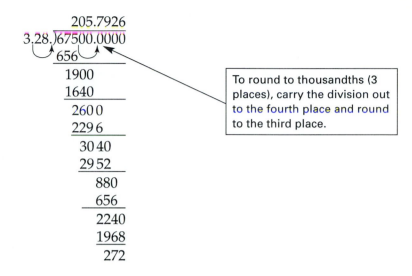

To round to thousandths (3 places), carry the division out to the fourth place and round to the third place.

Answer: 205.7926 = 205.793

Work Space

unit 4.7

Exercises

Divide each of these problems. Where needed, round off to the nearest thousandth (3 places):

ANSWERS

1. 8)‾.248

2. 7)‾.286

3. 5)‾2.45

4. 12)‾4.88

5. 9)‾24.5

6. 3)‾.033

7. 8)‾2.49

8. 2)‾.0012

9. 5)‾.028

10. 7)‾34.765

11. 15)‾.453

12. 21)‾.4221

1. _____
2. _____
3. _____
4. _____
5. _____
6. _____
7. _____
8. _____
9. _____
10. _____
11. _____
12. _____

Divide each of these problems. Give answers in dollar amounts rounded to the nearest cent (nearest hundredth):

13. $112.50 ÷ 15

14. $1,245.18 ÷ 12

13. _____

14. _____

15. _____

16. _____

17. _____

18. _____

19. _____

20. _____

15. $3,456.82 ÷ 25

16. $5,693.57 ÷ 8

17. $2,458.24 ÷ 14

18. $12,345.15 ÷ 212

19. $24,568.12 ÷ 240

20. $32,618.40 ÷ 30

Divide each of these problems. Where necessary, round answers to the nearest hundredth (2 places):

21. $.02\overline{).45}$

22. $.08\overline{)1.25}$

23. $.5\overline{)25}$

24. $1.2\overline{)12.78}$

25. $.002\overline{)8}$

26. $.045\overline{)4.5}$

27. $.9\overline{)81}$

28. $.08\overline{)6.4}$

29. $.007\overline{).49}$

30. $.25\overline{)26.75}$

31. $.09\overline{)27.29}$

32. $3.5\overline{)142.3}$

33. $.045\overline{)345.6}$

34. $.40\overline{)5.82}$

35. $1.4\overline{)50}$

21. _____

22. _____

23. _____

24. _____

25. _____

26. _____

27. _____

28. _____

29. _____

30. _____

31. _____

32. _____

33. _____

34. _____

35. _____

36. $.23\overline{)428}$ **37.** $.08\overline{)30.2}$ **38.** $5.25\overline{)75}$

ANSWERS

36. _____
37. _____
38. _____
39. _____
40. _____
41. _____
42. _____
43. _____
44. _____
45. _____
46. _____
47. _____
48. _____
49. _____
50. _____

39. $.0002\overline{)1.2}$ **40.** $.5\overline{)235.6}$ **41.** $.075\overline{)12.4}$

42. $.51\overline{)221}$ **43.** $2.4\overline{)3.6}$ **44.** $.03\overline{)30}$

Divide these problems and round off to the nearest cent:

45. $25\overline{)\$525.95}$ **46.** $40\overline{)\$324.58}$ **47.** $30\overline{)\$95.80}$

48. $36\overline{)\$568.20}$ **49.** $52\overline{)\$454.98}$ **50.** $25\overline{)\$125.49}$

Name _____ Date_____

Applications

Where appropriate, round all answers to even cents (even hundredths).

1. The total cost of a new car, including finance charges, is $12,456.88. If this amount can be paid in 48 monthly installments, what is the amount of each payment?

2. The Brewer Wholesale Company had 500 product catalogs printed at a total cost of $26,250. What is the cost of each catalog?

3. Professor Samuel Moore drove 1,456.8 miles on a recent lecture tour. How many miles per gallon did his car average if he bought a total of 56 gallons of gas?

4. Mendocin Company paid $945.52 to have a conference room carpeted. What was the cost per square yard if the area of the conference room is 50 square yards?

5. What is the cost of an orange if a dozen oranges cost $1.89?

6. Otto paid $13.86 to have his gas tank filled. How many gallons of gas did he purchase if one gallon costs $.99?

7. Franklin Hobbs earned $220 last week. If he earns $5.50 an hour, how many hours did he work?

1. _____

2. _____

3. _____

4. _____

5. _____

6. _____

7. _____

8. As part of a profit sharing plan, Video Products, Inc. is going to distribute $12,450.80 to its 300 employees. If each employee is to get an equal amount, what is the share for each employee?

9. Margaret Komendantov bought a stereo for $582 on time. How many payments will she make if one payment is $48.50?

10. If a gross (144) of writing pads costs $129.60, what is the cost per pad?

Work Space

unit 4.8

Dividing Decimals Using Shortcuts

After completing Unit 4.8, you will be able to:

1. Shortcut divide using a power of 10.
2. Shortcut divide using 25 and 50.

In Unit 4.6, you worked with some shortcuts that can help when multiplying decimal numbers. In this unit, we are going to look at some common shortcuts that help when dividing decimal numbers.

DIVIDING BY A POWER OF 10

To divide a decimal number by a power of 10 (10, 100, 1,000, etc.), move the decimal point in the dividend *to the left* the same number of places as there are zeros in the divisor.

Example 1
Divide 645.8 ÷ 10.

Solution
Move the decimal point one place to the left:

$$645.8 \div 10 = 645.8 = 64.58$$

Example 2
Divide: 56.45 ÷ 1,000.

Solution
Move the decimal point 3 places to the left:

$$56.45 \div 1,000 = 056.45 = .05645$$

Note: A zero was placed to the right of the decimal as a placeholder.

Example 3

Divide: $324 \div 100$.

Solution

Move the decimal point 2 places to the left:

$$324 \div 100 = 324. = 3.24$$

> The decimal point in a whole number is always after the right-hand digit.

Self-Check

Shortcut divide these numbers:

(a) 38.7 by 10   **(b)** 67.882 by 100.

Solution

(a) 3.87   **(b)** .67882

≋ DIVIDING BY 25 AND 50

To divide a number by 25, first multiply by 4, then divide by 100 (move decimal point 2 places to the left). For example, to divide 346.8 by 25:

Multiply by 4:

$$
\begin{array}{r}
346.8 \\
\times \quad 4 \\
\hline
1387.2
\end{array}
$$

Divide by 100:

$$1387.2 = \mathbf{13.872}$$

To divide a number by 50, first multiply by 2, then divide by 100. For example, to divide 468.42 by 50:

Multiply by 2:

$$
\begin{array}{r}
468.42 \\
\times \quad 2 \\
\hline
936.84
\end{array}
$$

Divide by 100:

$$936.84 = \mathbf{9.3684}$$

Remember:

1. To *divide* a number by a power of 10, move the decimal point in the dividend to the *left* the same number of places as there are zeros in the divisor. Be careful; if you move the decimal point to the *right*, you will be multiplying, not dividing.
2. To *divide* a number by 25, first multiply by 4, and then point off 2 places in the product.
3. To *divide* a number by 50, first multiply by 2, and then point off 2 places in the product.

Self-Check
Using shortcuts, divide these numbers:

(a) 800 by 25

(b) 4,000 by 50.

Solution
(a) 32

(b) 80

Work Space

Exercises

Name _____ Date _____

Divide each of the following problems in your head:

1. 345.8 ÷ 10

2. 48.2 ÷ 100

3. 2.7 ÷ 100

4. 124.89 ÷ 10

5. 1.2 ÷ 1,000

6. 15.45 ÷ 1,000

7. 2,458.2 ÷ 1,000

8. 3.48 ÷ 100

9. .07 ÷ 100

10. 4.64 ÷ 1,000

11. 500 ÷ 10,000

12. 89.1 ÷ 100

13. 3,456.99 ÷ 10,000

14. 318.009 ÷ 10

15. 8 ÷ 1,000

Divide each of these numbers by 10 in your head:

16. 80

17. 565

18. 3.45

19. 2.345

20. 565.8

21. 4,692

22. .4095

23. 748.775

ANSWERS

1. _____
2. _____
3. _____
4. _____
5. _____
6. _____
7. _____
8. _____
9. _____
10. _____
11. _____
12. _____
13. _____
14. _____
15. _____
16. _____
17. _____
18. _____
19. _____
20. _____
21. _____
22. _____
23. _____

Divide each of these numbers by 100 in your head:

24. 500

25. 658

26. 45.8

27. 238.65

28. .855

29. 12.5

30. 17.459

31. 675.4

Divide each of these numbers by 1,000 in your head:

32. 3,000

33. 33,500

34. 289.5

35. .8

36. 3,455.78

37. 401.89

38. 543.21

39. .055

Divide each of these numbers by 10,000 in your head:

40. 5,689

41. 4,000

42. 40,000

43. 12,345.89

44. 567.8

45. 178,563.76

46. 211.91

47. 3,456.72

Using shortcuts, divide each of these numbers by 25:

48. 400

49. 32.4

50. 12.82

51. 445.24

52. 1,632.44

53. 1,232.36

54. 6,416.48

55. 3,200

24. _____
25. _____
26. _____
27. _____
28. _____
29. _____
30. _____
31. _____
32. _____
33. _____
34. _____
35. _____
36. _____
37. _____
38. _____
39. _____
40. _____
41. _____
42. _____
43. _____
44. _____
45. _____
46. _____
47. _____
48. _____
49. _____
50. _____
51. _____
52. _____
53. _____
54. _____
55. _____

Using shortcuts, divide each of these numbers by 50:

56. 2,000

57. 4,200

58. 3,820.22

59. 240.84

60. 4,640.32

61. 2.008

62. 6,346.82

63. 12.78

56. _____

57. _____

58. _____

59. _____

60. _____

61. _____

62. _____

63. _____

Work Space

Name _____ **Date** _____

Applications

1. If 100 pens cost $56, what is the cost of one pen?

2. Steve Campione earned $87.50 in 10 hours. What is his wage per hour?

1. _____

2. _____

3. _____

4. _____

3. The Kissel Company's annual picnic and softball game cost $1,285. If this cost is to be shared equally by 100 employees, what is the cost per employee?

5. _____

6. _____

7. _____

4. West Concrete Company paid $12,485 for 10,000 advertising leaflets. To the nearest cent, what is the cost per leaflet?

5. The guests at a pool party sent out for pizza. The total cost of the pizza was $65, which was shared equally by 25 people. What is the cost per person?

6. If a box of thumb tacks that has 100 tacks costs $3.50, what is the cost of one tack?

7. Richard Marks paid $125 for snacks and gifts for a party for his fourth grade class. What was the cost per child if his class has a total of 50 students?

8. If 1,000 pounds of coffee has a wholesale price of $2,950, what is the cost of one pound?

9. On their last vacation trip, the Vella family traveled a total of 1,000 miles. What was their cost per mile if they spent $120 for gas?

10. Southland Foods had sales of $144,318 over a 100-day period. What was the average daily sales?

Work Space

unit 4.9

Equivalent Decimal and Common Fractions

After completing Unit 4.9, you will be able to:

1. Change a decimal fraction to an equivalent common fraction.
2. Change a common fraction to an equivalent decimal fraction.

Remember that a decimal is a special kind of fraction. Thus, all decimal fractions can be changed to equivalent common fractions, and all common fractions can be changed to equivalent decimal fractions. In business, common fractions are often changed to their decimal fraction equivalent to make problems easier to work with. Let's look at how this is done.

CHANGING A DECIMAL FRACTION TO AN EQUIVALENT COMMON FRACTION

An easy way to change a decimal fraction to a common fraction is to follow these steps:

Step 1: Read the decimal fraction aloud.

Step 2: Write the decimal as a common fraction.

Step 3: Reduce the common fraction to lowest terms, if needed.

Example 1
Change .125 to a common fraction in lowest terms.

Solution
Step 1: Read the decimal fraction: "one hundred twenty-five thousandths"

Step 2: Write the decimal as a common fraction:

$$\frac{125}{1,000}$$

Step 3: Reduce the common fraction to its lowest terms:

$$\frac{125}{1,000} = \frac{1}{8}$$

Example 2

Change .002 to a common fraction in lowest terms.

Solution

$$.002 = \frac{2}{1,000} = \frac{1}{500}$$

Example 3

Change 5.25 to a common fraction in lowest terms.

Solution

$$5.25 = 5\frac{25}{100} = 5\frac{1}{4}$$

Another way to change a decimal to a fraction is to count the number of places after the decimal point. Then, for your denominator, write the number 1 followed by as many zeros as there are places after the decimal point. For example:

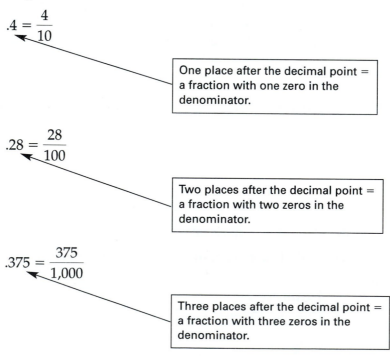

$$.4 = \frac{4}{10}$$

One place after the decimal point = a fraction with one zero in the denominator.

$$.28 = \frac{28}{100}$$

Two places after the decimal point = a fraction with two zeros in the denominator.

$$.375 = \frac{375}{1,000}$$

Three places after the decimal point = a fraction with three zeros in the denominator.

≋ CHANGING A COMMON FRACTION TO AN EQUIVALENT DECIMAL FRACTION

To change a common fraction to an equivalent decimal fraction, divide the denominator into the numerator. You may have to add zeros to the numerator to get the division to come out evenly.

Example 1
Change $\frac{5}{8}$ to an equivalent decimal fraction.

Solution

$$\frac{5}{8} = 8\overline{)5.000}$$

$$\begin{array}{r} .625 \\ 8\overline{)5.000} \\ \underline{4\,8} \\ 20 \\ \underline{16} \\ 40 \\ \underline{40} \end{array}$$

We added three zeros to make the division come out evenly.

Example 2
Change $\frac{1}{2}$ to an equivalent decimal fraction.

Solution

$$\frac{1}{2} = 2\overline{)1.0}$$

$$\begin{array}{r} .5 \\ 2\overline{)1.0} \\ \underline{1\,0} \end{array}$$

We added one zero to make the division come out evenly.

For numbers that cannot be divided evenly, you normally carry the division correct to 3 places. For example, when you change $\frac{2}{3}$ to a decimal, you will continue to get a remainder no matter how many zeros you add:

$$\frac{2}{3} = 3\overline{)\begin{array}{l} .666... \\ 2.000 \end{array}}$$

$$\begin{array}{r} \underline{1\ 8} \\ 20 \\ \underline{18} \\ 20 \\ \underline{18} \\ 2 \end{array}$$

> The three dots mean that the decimal will keep repeating, no matter how many zeros you add. It is read as *point 6 repeating*.

DECIMALS AND FRACTIONS IN A SINGLE PROBLEM

If you are working with a problem that contains both decimal fractions and common fractions, you must change the problem so that all fractions will be stated as either common fractions or decimal fractions.

Example 1
Add: $112.5 + 15\frac{3}{4} + 10\frac{1}{8}$

Solution
As a common fraction:
Change the decimal fraction to a common fraction.

$$\begin{array}{lll} 112.5 & 112\frac{1}{2} = & 112\frac{4}{8} \\ 15\frac{3}{4} = & 15\frac{3}{4} = & 15\frac{6}{8} \\ 10\frac{1}{8} & 10\frac{1}{8} = & 10\frac{1}{8} \\ & & \overline{137\frac{11}{8} = 138\frac{3}{8}} \end{array}$$

As a decimal fraction:
Change the common fraction to a decimal fraction.

$$\begin{array}{ll} 112.5 & 112.500 \\ 15\frac{3}{4} = & 15.750 \\ 10\frac{1}{8} & 10.125 \\ & \overline{138.375} \end{array}$$

Example 2
Multiply: $3.25 \times 2\frac{1}{2}$.

Solution
As a common fraction:

$$3\frac{1}{4} \times 2\frac{1}{2} = \frac{13}{4} \times \frac{5}{2} = \frac{65}{8} = 8\frac{1}{8}$$

As a decimal fraction:

$$\begin{array}{r} 3.25 \\ \times \quad 2.5 \\ \hline 1625 \\ 650 \quad \\ \hline 8125 = 8.125 \end{array}$$

Self-Check

Change $\frac{4}{15}$ to an equivalent decimal fraction. Round off correctly to the third place.

Solution

.267

Exercises

Name _____ Date_____

Change these decimal fractions to common fractions in their lowest terms:

1. .12 =

2. .015 =

3. .08 =

4. .002 =

5. .375 =

6. .025 =

7. .0025 =

8. .3 =

9. .03 =

10. .0075 =

11. .60 =

12. .425 =

13. .1245 =

14. .90 =

15. .00125 =

16. .45 =

17. .005 =

18. .0085 =

19. .625 =

20. .2 =

21. .09 =

ANSWERS

1. _____

2. _____

3. _____

4. _____

5. _____

6. _____

7. _____

8. _____

9. _____

10. _____

11. _____

12. _____

13. _____

14. _____

15. _____

16. _____

17. _____

18. _____

19. _____

20. _____

21. _____

Change these common fractions to equivalent decimal fractions. Round off correctly to the third place where applicable.

22. $\dfrac{7}{8}$ 23. $\dfrac{3}{4}$ 24. $\dfrac{1}{3}$

25. $\dfrac{4}{5}$ 26. $\dfrac{5}{7}$ 27. $\dfrac{2}{3}$

28. $\dfrac{1}{7}$ 29. $\dfrac{11}{12}$ 30. $\dfrac{1}{20}$

31. $\dfrac{5}{6}$ 32. $\dfrac{1}{9}$ 33. $\dfrac{1}{8}$

34. $\dfrac{1}{6}$ 35. $\dfrac{4}{9}$ 36. $\dfrac{7}{11}$

37. $\dfrac{1}{2}$ 38. $\dfrac{6}{7}$ 39. $\dfrac{5}{12}$

40. $\dfrac{17}{18}$ 41. $\dfrac{15}{19}$

22. _____
23. _____
24. _____
25. _____
26. _____
27. _____
28. _____
29. _____
30. _____
31. _____
32. _____
33. _____
34. _____
35. _____
36. _____
37. _____
38. _____
39. _____
40. _____
41. _____

Give the equivalent fraction and decimal answers to each of the following.
Round answers to 3 places where applicable:

	Fractional Method	**Decimal Method**
42. $12\frac{1}{8} + 8\frac{3}{4}$	_____	_____
43. $125.55 + 2\frac{3}{8} + 1\frac{4}{5}$	_____	_____
44. $312\frac{7}{8} + 225.75$	_____	_____
45. $29.25 - 3\frac{4}{5}$	_____	_____
46. $315\frac{1}{8} - 15.4$	_____	_____

	Fractional Method	**Decimal Method**
47. $3\frac{4}{5} \times 2.8$	_____	_____
48. $12.2 \times 10\frac{3}{5}$	_____	_____
49. $16.4 \div 4\frac{2}{5}$	_____	_____
50. $40\frac{5}{8} \div 2\frac{1}{10}$	_____	_____

unit 4.9

Applications

1. American Steel Builders bought $69\frac{1}{2}$ pounds of cleaning compound at a cost of $3.50 a pound. What was the total cost?

2. What is the cost of $32\frac{1}{2}$ square yards of carpet if one square yard costs $8.40?

3. A pack of sandwich style turkey breast weighs $\frac{9}{16}$ of a pound. What is its weight to the nearest *hundredth* of a pound?

4. Dave jogs $\frac{7}{8}$ of a mile each day on his lunch hour. How many *hundredths* of a mile does he jog?

5. A certain headache medicine is sold in packets that weigh one ounce. How many *thousandths* of a pound does it weigh? (*Hint:* one ounce $= \frac{1}{16}$ of a pound.)

ANSWERS

1. _____

2. _____

3. _____

4. _____

5. _____

6. A $100 leather coat is advertised as $\frac{1}{4}$ off. How many cents on the dollar can be saved by buying the coat at this special price?

7. Last week Dennis Yarrow worked $38\frac{1}{4}$ hours. What are his total earnings if he earns $7.45 an hour?

8. Which is larger: $\frac{7}{8}$ or $\frac{8}{9}$? (*Hint:* Convert the fractions to decimals and compare the decimals.)

9. To clean his carpets, Brad Brady needs $\frac{5}{6}$ of a gallon liquid soap. How many *hundredths* of a gallon does he need?

10. Burger Master bought $40\frac{3}{8}$ pounds of ground beef at a cost of $.99 per pound. What was the total cost?

section

V

Percents

unit 5.1

Understanding Percents

After completing Unit 5.1, you will be able to:

1. Change decimals to percents.
2. Change percents to decimals.
3. Change fractions to percents.
4. Change percents to fractions.

In earlier units, we studied two ways of expressing a part of a whole unit—decimals and fractions. If, for example, you want to express "one part out of a total of four parts," the fraction $\frac{1}{4}$ will do this. The decimal .25 (twenty-five hundredths) can also be used to express one part of a total of four parts. Thus, $\frac{1}{4}$ and .25 are different ways of expressing the same value.

A third way to express a part of a whole unit is through **percent,** which means *by the hundred* or **out of a hundred.** Thus, **one percent** is "one part out of a total of 100 parts," and **10 percent** is "10 parts out of a total of 100 parts." A dollar, for example, has 100 cents; thus, 10 cents equals 10 percent of a dollar.

≋ A PERCENT IS A FRACTION WITH A DENOMINATOR OF 100

A percent is like a decimal in that they are both fractions with an understood denominator. The denominator of a decimal fraction, as you recall, is always 10, 100, 1,000, or some other power of 10; the denominator of a percent is always 100.

When working with percents, we often use the percent sign (%) in place of the word "percent." Thus, 15% is read as "15 percent," and $12\frac{1}{2}\%$ is read as "twelve and one-half percent."

Because decimals, percents, and fractions are all different ways of expressing the same values, we can easily change one form of expression to another. We will see how to do this next.

CHANGING DECIMALS TO PERCENTS

To change a decimal to a percent, move the decimal point 2 places to the right and add the percent sign.

Examples

Change the following decimals to percents:

.12
.08
.375

Solution

.12 = .12 = 12%

.08 = .08 = 8%

.375 = .375 = 37.5% (or 37½%)

Whole numbers and mixed numbers are changed to percents in the same way:

5 = 5.00 = 500%

2 = 2.00 = 200%

3.5 = 3.50 = 350%

Self-Check

Change .875 to a percent.

Solution

87.5% (87½%)

CHANGING PERCENTS TO DECIMALS

To change a percent to a decimal, reverse the preceding operation. That is, you move the decimal point 2 places to the left and drop the percent sign.

Examples

Change the following percents to decimals:

18%
6%
67.5%
2.5%

Solution

$$18\% = .18$$

$$06\% = .06$$

$$67.5\% = .675$$

$$02.5\% = .025$$

Self-Check

Change 93.8% to a decimal.

Solution

.938

CHANGING FRACTIONS TO PERCENTS

To change a fraction to a percent, first change the fraction to an equivalent decimal; then move the decimal point 2 places to the right and add on the percent sign.

Examples

Change the following fractions to percents:

$$\frac{1}{2}$$

$$\frac{3}{4}$$

$$2\frac{1}{4}$$

Solution

Change the fractions to decimals, then change the decimals to percents:

Fraction ⟶ *Decimals* ⟶ *Percents*

$$\frac{1}{2} = 1 \div 2 = \rightarrow \qquad .5 = .50 = \rightarrow \qquad 50\%$$

$$\frac{3}{4} = 3 \div 4 = \rightarrow \qquad .75 = .75 = \rightarrow \qquad 75\%$$

$$2\frac{1}{4} = 9 \div 4 = \rightarrow \qquad 2.25 = 2.25 = \rightarrow \qquad 225\%$$

Remember: $2\frac{1}{4} = \dfrac{(4 \times 2 = 8) + 1}{4} = \dfrac{9}{4}$

Look at the last example; another way to change $2\frac{1}{4}$ to 225% is as follows:

$$2\frac{1}{4} = 2.25 = 2.25 = 225\%$$

Self-Check

Change 2% to a fraction.

Solution

$\frac{1}{50}$

CHANGING PERCENTS TO FRACTIONS

To change a percent to a fraction, first change the percent to a decimal; then change the decimal to a fraction. Reduce the fraction to its lowest terms, if needed.

Examples

Change the following percents to fractions:

$$4\%$$
$$25\%$$
$$80\%$$
$$2.75\%$$

Solution

Change the percent to a decimal, then change the decimal to a fraction:

Percent ⟶ *Decimal* ⟶ *Fraction*

$$4\% = .04\% = \rightarrow \qquad .04 = \rightarrow \qquad \frac{4}{100} = \frac{1}{25}$$

$$25\% = .25\% = \rightarrow \qquad .25 = \rightarrow \qquad \frac{25}{100} = \frac{1}{4}$$

$$80\% = .80\% = \rightarrow \qquad .80 = \rightarrow \qquad \frac{80}{100} = \frac{4}{5}$$

$$2.75\% = .0275\% = \rightarrow \qquad .0275 = \rightarrow \qquad \frac{275}{10,000} = \frac{11}{400}$$

Another way to change a percent to a fraction is to remember that a percent is a fraction with a denominator of 100:

$$5\% = \frac{5}{100} = \frac{1}{20}$$

$$40\% = \frac{40}{100} = \frac{2}{5}$$

$$55\% = \frac{55}{100} = \frac{11}{20}$$

Remember:

1. Percent means "by the hundred."
2. A percent is a special type of fraction that always has an understood denominator of 100:

$$50\% = \frac{50}{100}$$

$$6\% = \frac{6}{100}$$

$$33\% = \frac{33}{100}$$

3. Since fractions, decimals, and percents are different ways of expressing the same values, you can easily change from one form of expression to another.

Work Space

unit
5.1

Exercises

Fill in the following blanks:

1. $\dfrac{10}{100} =$ _____%

2. $\dfrac{6}{100} =$ _____%

3. $\dfrac{85}{100} =$ _____%

4. $\dfrac{5}{100} =$ _____%

5. $\dfrac{7}{100} =$ _____%

6. $\dfrac{125}{100} =$ _____%

7. $\dfrac{2}{100} =$ _____%

8. $\dfrac{375}{100} =$ _____%

9. $\dfrac{72.5}{100} =$ _____%

10. 45 out of 100 = _____%

11. 31 out of 100 = _____%

12. $2\frac{1}{2}$ out of 100 = _____%

13. 99 out of 100 = _____%

14. $12\frac{1}{3}$ out of 100 = _____%

15. 1 out of 100 = _____%

16. 4 hundredths = _____%

17. 35 hundredths = _____%

18. 225 hundredths = _____%

19. 17 hundredths = _____%

20. 12.5 hundredths = _____%

21. 100 hundredths = _____%

Change each of these decimals to percents:

22. .45 =   **23.** .98 =   **24.** 1.05 =   **25.** 2.75 =

26. .375 =   **27.** .17 =   **28.** 1.24 =   **29.** .02 =

30. 3.50 =   **31.** 5 =   **32.** .376 =   **33.** .50 =

Change each of these percents to decimals:

34. 5% =   **35.** 8% =   **36.** 12% =

37. 125% =   **38.** 28.5% =   **39.** 200% =

40. 34.8% =   **41.** 2.5% =   **42.** 100% =

Change each of these fractions to percents:

43. $\dfrac{3}{4}$ =   **44.** $\dfrac{1}{2}$ =   **45.** $\dfrac{4}{5}$ =

46. $\dfrac{3}{5}$ =   **47.** $2\dfrac{1}{2}$ =   **48.** $\dfrac{1}{4}$ =

49. $4\dfrac{3}{4}$ =   **50.** $\dfrac{5}{6}$ =   **51.** $\dfrac{1}{8}$ =

22. _____
23. _____
24. _____
25. _____
26. _____
27. _____
28. _____
29. _____
30. _____
31. _____
32. _____
33. _____
34. _____
35. _____
36. _____
37. _____
38. _____
39. _____
40. _____
41. _____
42. _____
43. _____
44. _____
45. _____
46. _____
47. _____
48. _____
49. _____
50. _____
51. _____

Change each of these percents to fractions. Reduce to their lowest terms where needed:

52. 24% =

53. 45% =

54. 18% =

55. 15% =

56. $2\frac{1}{2}\% =$

57. $6\frac{1}{4}\% =$

58. 225% =

59. 5.8% =

60. .5% =

52. _____

53. _____

54. _____

55. _____

56. _____

57. _____

58. _____

59. _____

60. _____

Complete the following:

	Fraction	Decimal	Percent
61.	$\frac{4}{5}$	_____	_____
62.	_____	.80	_____
63.	_____	_____	10%
64.	$\frac{2}{3}$	_____	_____
65.	_____	.025	_____
66.	_____	_____	125%
67.	_____	.95	_____
68.	$\frac{9}{10}$	_____	_____

Work Space

Name _____ Date _____

Applications

1. Of the workers at MUTECH Company, 52% are women. For each 100 workers, how many are women?

2. One Monday morning 20% of the students at Eastdale High were absent. What percent of the student body was present?

3. If a suit is marked down by 30%, what percent of the selling price will be paid?

4. Glenn Seymour received a 25% raise in pay. What percent is his new salary compared to his old salary?

5. If 71% of the employees at Denver Company have health insurance, what percent do not?

6. In a recent election, there was a 90% voter turnout. What percent of voters did not vote?

7. Jane McKay is in the 40% tax bracket. What percent of her pay does she take home?

ANSWERS

1. _____
2. _____
3. _____
4. _____
5. _____
6. _____
7. _____

8. What percent of a dollar is 75 cents?

ANSWERS

8. _____

9. _____

10. _____

9. In a certain Florida town, eight out of ten people are retired. Express this as a percent.

10. What is 25% of a dollar?

Work Space

unit 5.2

Working with the Base, Rate, and Portion

After completing Unit 5.2, you will be able to:

1. Identify and find the portion.
2. Identify and find the rate.
3. Identify and find the base.

To start our discussion, let's look at a transaction that is typical of the many thousands of business transactions that occur each day in our country.

John Dorman is a real estate salesperson. Last week John sold a house for $90,000. His 6% commission for selling the house was $5,400.

This application of percent identifies three elements that you should be very familiar with: the base, rate, and portion.

Base. The **base** is the whole value, or the beginning quantity. In the preceding example, the base is the price of the house, $90,000. The word "of" or "part of" is often written before the base. Thus, writing the problem as "6% *of* $90,000 equals the commission on the sale" would help to identify the $90,000 as the base in this problem.

Rate. The **rate** is a percent (such as 8%, 10%, $12\frac{1}{2}$%, etc.). The rate indicates the part of the base that is to be calculated. In the example, the rate is 6%. As a general rule, the rate can be easily identified because it is usually written with the percent sign.

Portion. The **portion** is the result of multiplying the base by the rate. In the example, the portion is $5,400, which resulted from multiplying the base, $90,000, by the rate, 6% ($90,000 $\times$ 6% = $5,400).

We have now identified the three elements of a percent problem: the base, rate, and portion. If two of these elements are known, the third can always be calculated. Let's look at how to find a missing element when the other two are given.

NOTE
The portion is sometimes called the "percentage." We use the term "portion" because it is easy to confuse the terms "percent" and "percentage."

≋ FINDING THE PORTION (P)

Since the portion is the product of the base and the rate, we can use this formula to find the portion:

$$\text{Base} \times \text{rate} = \text{portion}$$
or
$$B \times R = P$$

Example 1

If the base is 800 and the rate is 8%, find the portion.

Solution

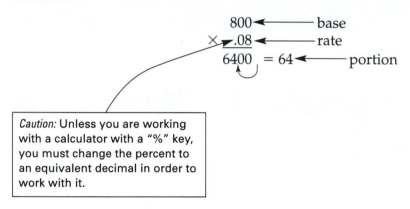

$$
\begin{array}{r}
800 \xleftarrow{\hspace{2cm}} \text{base} \\
\times \ .08 \xleftarrow{\hspace{2cm}} \text{rate} \\
\hline
6400 = 64 \xleftarrow{\hspace{2cm}} \text{portion}
\end{array}
$$

Caution: Unless you are working with a calculator with a "%" key, you must change the percent to an equivalent decimal in order to work with it.

Example 2

If the base is $400 and the rate is $2\frac{1}{2}\%$, find the portion.

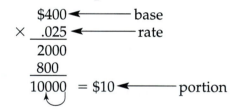

$$
\begin{array}{r}
\$400 \xleftarrow{\hspace{2cm}} \text{base} \\
\times \ .025 \xleftarrow{\hspace{2cm}} \text{rate} \\
\hline
2000 \\
800 \ \\
\hline
10000 = \$10 \xleftarrow{\hspace{2cm}} \text{portion}
\end{array}
$$

Example 3

Find 25% of 600.

Solution

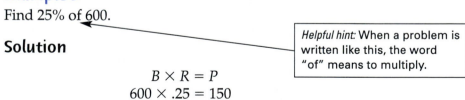

Helpful hint: When a problem is written like this, the word "of" means to multiply.

$$B \times R = P$$
$$600 \times .25 = 150$$

In some problems, you will need to find the portion even though the problem does not directly say so.

Example

Jim Pearson earns $900 a month. He spends 20% of this amount for rent. Determine his rent expense.

Solution

Base = $900 (Remember that the base is the whole quantity.)
Rate = 20% (Remember that the "%" mark will usually identify the rate.)
Portion = ?

$$B \times R = P$$
$$\$900 \times .20 = \$180$$

When you solve percent problems, often it helps to use a pie diagram, like this:

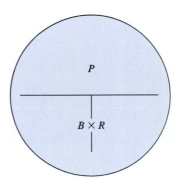

To use the pie diagram, place your finger over the missing element; the elements shown will then show you how to find the missing element. To find the portion, for example, place a finger over the P:

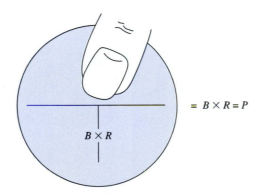

$= B \times R = P$

Example

Using a pie diagram, find the portion when the base is 500 and the rate is 12%.

Solution

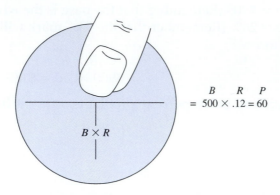

$$\begin{array}{ccc} B & R & P \\ = 500 \times .12 = 60 \end{array}$$

$B \times R$

Self-Check

The base is $8,000 and the rate is 12%. Find the portion.

Solution

$960

≋ FINDING THE RATE (R)

The rate can be found as follows:

$$P \div B = R$$

Example

The portion is $54 and the base is 900. What is the rate?

Solution

$$P \div B = R$$
$$\$54 \div \$900 = 6\%$$

$$\begin{array}{r} .06 = 6\% \\ 900\overline{)54.00} \\ 54\ 00 \end{array}$$

The use of a pie diagram will also help when looking for the rate. For example, let's solve the preceding example with the use of a pie diagram.

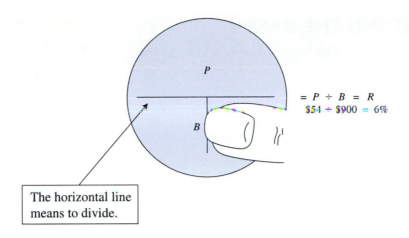

$= P \div B = R$
$\$54 \div \$900 = 6\%$

The horizontal line means to divide.

Let's look at another example.

Example

Dean Watson received a \$32.50 commission for selling \$650 worth of merchandise. What was his rate of commission?

Solution

$B = \$650$ (Remember that the base is the beginning quantity.)
$P = \$32.50$ (the part of the whole)
$R = ?$

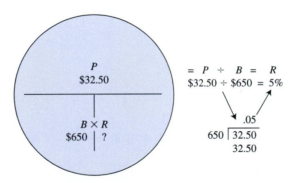

$= P \div B = R$
$\$32.50 \div \$650 = 5\%$

$$\begin{array}{r} .05 \\ 650\overline{\smash{\big)}\,32.50} \\ 32.50 \end{array}$$

Check: $\$650 \times .05 = \$32.50$

Self-Check

The base is \$6,200 and the portion is \$930. What is the rate?

Solution

15%

≋ FINDING THE BASE

The base can be found as follows:

$$P \div R = B$$

Example

The portion is $36 and the rate is 9%. Find the base.

Solution

$$P \div R = B$$
$$\$36 \div .09 = \$400$$

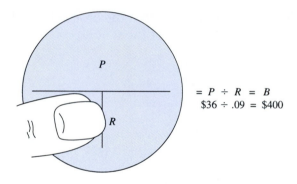

$$\begin{array}{r} 400. \\ .09\overline{)36,00.} \\ 3600 \end{array}$$

Check: $400 × .09 = $36

Again, we can use a pie diagram to help find the missing element—the base, in this case. Let's look again at the preceding example:

$$= P \div R = B$$
$$\$36 \div .09 = \$400$$

Let's look at another example where we need to calculate the base.

Example

Of what number is 8% equal to 72?

Solution

P = 72 (the part of the whole)
R = 8% (The "%" sign identifies the rate.)
B = ? (The beginning whole quantity is missing.)

$$P \div R = B$$
$$72 \div .08 = 900$$

Check: 900 × .08 = 72

REMEMBER

$B × R = P$
$P \div B = R$
$P \div R = B$

The portion is 820 and the rate is 8%. Find the base.

Solution

10,250

Work Space

Name _____ Date _____

Exercises

Find the portion in each of these problems. Remember to change the rates to decimals before you multiply:

1. 5% of 48

2. 20% of 24

3. 75% of 100

4. 8% of 500

5. 90% of 2,000

6. 42% of 300

7. 10% of 50

8. 16% of 250

9. 11% of 400

10. 60% of $3,500

11. 55% of $350

12. 37% of $1,200

13. 14% of $525

14. 28% of $960

15. 25% of $1,650

16. 95% of $585

17. 6% of $35,000

18. 9% of $45,600

ANSWERS

1. _____
2. _____
3. _____
4. _____
5. _____
6. _____
7. _____
8. _____
9. _____
10. _____
11. _____
12. _____
13. _____
14. _____
15. _____
16. _____
17. _____
18. _____

Find the rate in each of these problems:

	Base	Portion	Rate
19.	$ 6,000	$ 600	_____%
20.	8,500	1,020	_____
21.	2,450	490	_____
22.	9,000	450	_____
23.	12,250	9,800	_____
24.	18,625	5,960	_____
25.	4,000	2,000	_____
26.	12,000	240	_____
27.	1,500	105	_____

Find the base in each of these problems:

	Portion	Rate	Base
28.	300	10%	_____
29.	225	8%	_____
30.	630	12%	_____
31.	4,500	9%	_____
32.	3,600	6%	_____
33.	900	3%	_____
34.	450	5%	_____
35.	1,800	9%	_____
36.	7,200	8%	_____

Find the missing element in each of these problems:

Base	Rate	Portion
37. $30,000.00	15%	$_____
38. $ 6,800.00	_____	$ 272.00
39. _____	40%	$6,000.00
40. $ 455.80	8%	_____
41. _____	6%	$ 735.30
42. $12,890.75	_____	$1,289.08

Work Space

unit 5.2

Name _____ **Date** _____

Applications

1. Identify the base, rate, and portion in each of the following statements:

   a. 25% of 300 is 75

   b. $200 is 20% of $1,000

   c. $6,000 × .08 = $480

2. What word, or words, normally indicate the base in a problem?

3. Jane Murrie is a sales representative for a large grocery company. Jane is paid a 6% commission on all goods she sells. What is her commission in a month in which her total sales amounted to $55,250?

4. Jim McGuire pays 30% of his earnings in taxes. What are his taxes in a week in which he earned $350?

5. On his last payday, Robert Santaliz deposited $60 in his savings account. If the deposit was 5% of his total earnings, what were his total earnings?

6. Reda Kinza, a real estate salesperson, sold a house for $75,000. What is the amount of her commission if her rate of commission is 7%?

ANSWERS

1. a. _____

_____

_____

b. _____

_____

_____

c. _____

_____

_____

2. _____
3. _____
4. _____
5. _____
6. _____

7. Jeff Boling sold a car at a profit of $400. If his profit was 40% of what he paid for the car, how much did the car cost him?

8. At a local school, 250 students are taking a bookkeeping course. If this is 10% of the total school enrollment, how many students does the school have?

9. Jenna Rubin paid a 6% sales tax on the purchase of a $12,000 car. What is (a) the amount of sales tax and (b) the total cost of the car after the sales tax is added?

10. An employment test given by a telephone company was passed by 35% of the people who took the test. How many passed if a total of 500 people took the test?

11. Last year the operating expenses of SmartTech Company were 40% of the company's sales revenue. If the operating expenses were $360,000, what were the total sales?

12. After payroll taxes are withheld, the amount of Ray Kinney's weekly payroll check is $180. What rate of taxes is he paying if he earns $250 before taxes?

unit
5.3

Additional Applications of Percents

After completing Unit 5.3, you will be able to:

1. Find the percent of one number to another.
2. Find the rate of increase or decrease in two numbers.

In the preceding two units, we reviewed percents and studied their application to common business situations. In this unit, we are going to look at some additional topics in percent.

≋ FRACTIONAL PARTS OF 1%

We have learned that 1% means "1 part out of 100 parts." In other words, 1% is a fraction of 100%. But, sometimes percents are expressed as fractions of 1%. For example, $\frac{1}{4}$% means "$\frac{1}{4}$ of 1%" and $\frac{2}{3}$% means "$\frac{2}{3}$ of 1%." Such expressions are fractions of 1%, *not fractions of 100%*. So, you should be careful when working with them.

When working with fractions of 1%, it is usually easier to change them to their decimal equivalents, which we do as follows:

To change a fraction of 1% to its decimal equivalent, you:

Step 1: Rewrite the common fraction as a decimal fraction.

Step 2: Drop the "%" sign and move the decimal point 2 places to the left.

Example 1

Change $\frac{1}{4}\%$ to a decimal.

Solution

Step 1: Rewrite the common fraction as a decimal:

$$\frac{1}{4}\% = .25\%$$

Since $\frac{1}{4} = .25$, $\frac{1}{4}\% = .25\%$

Step 2: Drop the "%" sign and move the decimal point 2 places to the left: ⟶ But w/The % sign, it Remains .25%.

$$.25\% = 00.25\% = .0025$$

Example 2

Change $\frac{3}{8}\%$ to a decimal.

Solution

$$\frac{3}{8}\% = .375\% = .00.375\% = .00375$$

Since $\frac{3}{8} = .375$, $\frac{3}{8}\% = .375\%$

As we have learned, some fractions do not come out evenly when they are changed to decimal equivalents. When this happens, you must round off the decimal to a given place.

Example

Change $\frac{2}{3}\%$ to a decimal rounded correctly to 3 places.

Solution

$$\frac{2}{3}\% = 3\overline{)2.0000}^{\,.6666} = .667\% = .00.667\% = .00667$$

Remember: To round a decimal correct to 3 places, you must carry out the decimal to 4 places and round back to 3.

〰 FINDING THE PERCENT OF ONE NUMBER TO ANOTHER NUMBER

In business, often you have to find the percent of one number to another number. To do this, start out by writing the numbers as a fraction. You then change the fraction to a percent.

Example 1

What percent of 80 is 40?

Solution

Express 40 as a fraction of 80:

$$\frac{40}{80} = \frac{1}{2}$$

Change the fraction to a percent:

$$\frac{1}{2} = .50 = 50\%$$

Answer = 50%

Another way to solve this problem is by using the formula for finding the rate (studied in Unit 5.2):

$P \div B = R$
$40 \div 80 = 50\%$

$$\frac{.50}{80)\overline{40.00}} \qquad .50 = .50 = 50\%$$

Example 2

Of the 200 workers at Arcon Industries, 25 were absent on Friday. Find the percent of those absent.

Solution

$$\frac{25}{200} = \frac{1}{8} = .125 = .125 = 12.5\% \text{ (or } 12\frac{1}{2}\%)$$

or

$$P \div B = R$$
$$25 \div 200 = 12.5\%$$

> Remember: The base is the beginning whole quantity.

> The portion is the part of the base.

Self-Check

What percent of 90 is 30?

Solution

$33\frac{1}{3}\%$

≈≈≈ FINDING THE PERCENT OF INCREASE OR DECREASE

Finding the percent of increase or decrease in amounts is a very common business application. For example, the manager of Lakeside Grocery noticed that produce sales rose from $16,000 in May to $20,000 in June. To compare the increase with other months, he calculated the percent of increase (or the rate of increase). To find the percent of increase (or decrease), first find the amount of increase or decrease; you then divide the amount of increase or decrease by the original amount. We can state this in formula form, as follows:

$$\text{Percent of increase or decrease} = \frac{\text{amount of increase or decrease}}{\text{original amount}}$$

Example 1

If Lakeside Grocery's produce sales went from $16,000 in May to $20,000 in June, find the percent of increase.

Solution

Find the amount of increase:

$20,000 ◄——————— new amount
− 16,000 ◄——————— original amount
$ 4,000 ◄——————— amount of increase

Divide the amount of increase by the original amount:

$$\frac{.25}{\$16,000)\$4,000.00} = \textbf{25\% increase}$$

Example 2

When MicroTech got a new computer system, the company was able to reduce its operating expenses from $80,000 in 20X1 to $72,800 in 20X2. What was the percent decrease?

Solution

$80,000 ◄——————— original amount
− 72,800 ◄——————— new amount
$ 7,200 ◄——————— amount of change

$$\frac{.09}{\$80,000)\$7,200.00} = \textbf{9\% decrease}$$

Self-Check

The price of movie tickets in New York City recently increased from $8.00 to $8.50. What was the rate of increase?

Solution

6.25% ($6\frac{1}{4}$%)

Name _____ Date_____

Exercises

Change each of these to a decimal. Where necessary, round correctly to 3 places:

ANSWERS

1. $\dfrac{1}{2}\% =$

2. $\dfrac{5}{8}\% =$

3. $\dfrac{1}{10}\% =$

4. $\dfrac{3}{8}\% =$

5. $\dfrac{7}{10}\% =$

6. $\dfrac{4}{5}\% =$

7. $\dfrac{1}{5}\% =$

8. $\dfrac{7}{8}\% =$

9. $\dfrac{1}{3}\% =$

10. $\dfrac{3}{10}\% =$

11. $\dfrac{2}{5}\% =$

12. $\dfrac{2}{15}\% =$

13. $\dfrac{18}{25}\% =$

14. $\dfrac{9}{15}\% =$

1. _____
2. _____
3. _____
4. _____
5. _____
6. _____
7. _____
8. _____
9. _____
10. _____
11. _____
12. _____
13. _____
14. _____

Complete the following. Where applicable, round final answers correctly to 2 places:

15. $\frac{1}{4}$% of 80 =

16. $\frac{1}{5}$% of $200 =

17. $\frac{3}{5}$% of $152.78 =

18. $\frac{9}{10}$% of $2,000 =

19. $\frac{4}{5}$% of 5,280 =

20. $\frac{1}{2}$% of $424.80 =

21. $\frac{3}{10}$% of $568.42 =

22. $\frac{2}{5}$% of 500 =

23. $\frac{3}{15}$% of $900 =

24. $\frac{3}{5}$% of $242.50 =

25. $\frac{3}{4}$% of 1,200 =

15. _____
16. _____
17. _____
18. _____
19. _____
20. _____
21. _____
22. _____
23. _____
24. _____
25. _____

Solve these problems. Round to the nearest whole percent:

26. 25 is _____% of 50

27. 32 is _____% of 80

28. 75 is _____% of 200

29. 45 is _____% of 135

30. 42 is _____% of 90

31. 200 is _____% of 1,200

32. 320 is _____% of 950

33. 125 is _____% of 500

34. 500 is _____% of 2,500

35. 450 is _____% of 1,400

State the following as the nearest whole percent:

36. 15 out of 45 =

37. 20 out of 80 =

38. 60 out of 90 =

39. 45 out of 120 =

40. 75 out of 300 =

41. 200 out of 600 =

42. 50 out of 800 =

43. 8 out of 200 =

44. 60 out of 420 =

45. 50 out of 225 =

ANSWERS

36. _____

37. _____

38. _____

39. _____

40. _____

41. _____

42. _____

43. _____

44. _____

45. _____

Find the rate of increase (or decrease) in each of the following. Where applicable, round percents to the nearest tenth percent:

New Amount	Original Amount	Amount of Change	Percent Change
46. $500	$ 300	$_____	_____%
47. 1,800	2,000	_____	_____%
48. 2,400	1,600	_____	_____%
49. 6,000	4,000	_____	_____%
50. 7,200	8,000	_____	_____%
51. 2,500	1,000	_____	_____%
52. 4,500	9,000	_____	_____%
53. 9,350	8,120	_____	_____%
54. 1,400	800	_____	_____%
55. 6,290	8,410	_____	_____%

Work Space

Applications

Name _____ Date_____

1. The population of Baldwin County grew "one-half of one percent" during June 19X6. If the population was 2,456,800 at the beginning of June, what is the population at the end of June?

2. Last month the economy's inflation rate was "nine-tenths of one percent." Write this as (a) a decimal and (b) as a percent.

3. Using decimal equivalents, prove that .1% and $\frac{1}{10}$% have the same value.

4. Which is larger, .8% or $\frac{7}{8}$%? How much larger? (*Hint:* Convert the percents to decimals and compare the decimals.)

5. What is $\frac{3}{4}$% of $200?

6. Jim and Barbara Miller are looking for financing for a new home. They qualified for loans at two different banks. One bank quoted a rate of $10\frac{7}{8}$%, and the other bank quoted a rate of 10.8%. Which rate is more attractive?

7. Find $\frac{9}{10}$% of $2,560.54.

8. James DeFoe earns $900 a month. But, he only takes home $720. What percent of his earnings does he take home?

ANSWERS

1. _____

2. a. _____

   b. _____

3. _____

4. _____

5. _____

6. _____

7. _____

8. _____

9. Of the 600 employees at Safeway Company, 320 are female. What percent of the work force is female?

10. A suit that usually sells for $195 was marked down to $125. What is the percent reduction?

11. Sue Lang got 94 on her first math test and 85 on her second math test. What was her percent change?

12. Last month movie tickets went up from $4.50 a ticket to $5.00 a ticket. What is the percent increase?

13. Buying a $60,000 house requires a $12,000 down payment. Give the down payment as a percent.

14. Fill out this financial statement for the Bellows Company. Round rates to the nearest whole percent:

BELLOWS COMPANY Position Statement December 31, 20X2 and 20X3				
	20X3	**20X2**	**Change**	**Percent Change**
Assets				
Cash	$ 12,000	$ 10,000	$_____	_____%
Receivables	28,000	26,000	_____	_____%
Inventory	30,000	24,000	_____	_____%
Supplies	1,000	1,200	_____	_____%
Equipment	90,000	72,000	_____	_____%
Liabilities				
Accounts payable	$ 8,000	$ 6,000	$_____	_____%
Taxes payable	1,200	1,400	$_____	_____%
Equity				
Owner's equity	$151,800	$125,800	$_____	_____%

Aliquot Parts

After completing Unit 5.4, you will be able to:

1. Convert rates to aliquot parts of $1.00 and 100%.

When you work with problems involving percents or decimals, calculations are often easier if you use a math shortcut called aliquot parts. An **aliquot part** is any number (whole or mixed) that divides evenly into another number without leaving a remainder. For example, $.25 is an aliquot part of $1.00 because $.25 divides into $1.00 exactly four times. In the same way, 20 is an aliquot part of 60; 5 is an aliquot part of 25; 9 is an aliquot part of 27; and $12\frac{1}{2}$ is an aliquot part of 100.

To illustrate how the use of aliquot parts can make problem solving easier, let's look at a couple of calculations.

Example
What is the cost of 24 items if the cost per item is $.25?

Solution

Regulation Calculation

$$
\begin{array}{r}
24 \\
\times \ \underline{\$.25} \\
120 \\
\underline{48 } \\
\$6.00
\end{array}
$$

Use of Aliquot Part

$$\frac{24}{1} \times \$\frac{1}{4} = \$6$$

> Since $.25 divides into $1.00 exactly four times, it is $\frac{1}{4}$ of a $1.00.

As we can see from these calculations, multiplying by $.25 is the same as multiplying by $\$\frac{1}{4}$ because $.25 is one-fourth ($\frac{1}{4}$) of $1.00. But, it is much quicker to multiply by $\frac{1}{4}$.

Since use of an aliquot part will give you the same results as using its decimal equivalent, which should you use, the decimal or the aliquot part? The answer is simple—aliquot parts are best used when they make your calculations easier. Following is a table of the common aliquot parts of $1.00; you should learn these:

COMMON ALIQUOT PARTS OF $1.00

$$\$\frac{1}{2} = 50 \text{ cents}$$

$$\$\frac{1}{3} = 33\frac{1}{3} \text{ cents}$$

$$\$\frac{1}{4} = 25 \text{ cents}$$

$$\$\frac{1}{5} = 20 \text{ cents}$$

$$\$\frac{1}{6} = 16\frac{2}{3} \text{ cents}$$

$$\$\frac{1}{7} = 14\frac{2}{7} \text{ cents}$$

$$\$\frac{1}{8} = 12\frac{1}{2} \text{ cents}$$

$$\$\frac{1}{10} = 10 \text{ cents}$$

$$\$\frac{1}{12} = 8\frac{1}{3} \text{ cents}$$

$$\$\frac{1}{15} = 6\frac{2}{3} \text{ cents}$$

$$\$\frac{1}{16} = 6\frac{1}{4} \text{ cents}$$

$$\$\frac{1}{20} = 5 \text{ cents}$$

$$\$\frac{2}{3} = 66\frac{2}{3} \text{ cents}$$

$$\$\frac{3}{4} = 75 \text{ cents}$$

$$\$\frac{2}{5} = 40 \text{ cents}$$

$$\$\frac{3}{5} = 60 \text{ cents}$$

$$\$\frac{4}{5} = 80 \text{ cents}$$

$$\$\frac{5}{6} = 83\frac{1}{3} \text{ cents}$$

$$\$\frac{3}{8} = 37\frac{1}{2} \text{ cents}$$

$$\$\frac{7}{8} = 87\frac{1}{2} \text{ cents}$$

Let's look at some examples to see how the use of aliquot parts can make problem solving easier.

Example 1

What is the cost of 96 items if the cost per item is $12\frac{1}{2}$ cents?

Solution

$12\frac{1}{2}$ cents $= \$\dfrac{1}{8}$; so:

$$\dfrac{\overset{12}{\cancel{96}}}{1} \times \$\dfrac{1}{\cancel{8}} = \$12$$

Example 2

Calculate the total price of 64 pens if one pen costs $37\frac{1}{2}$ cents.

Solution

$37\frac{1}{2}$ cents $= \$\dfrac{3}{8}$

$$\dfrac{\overset{8}{\cancel{64}}}{1} \times \$\dfrac{3}{\cancel{8}} = \$24$$

Aliquot parts can also be used for values that are greater than $1.00. For example, what is the cost of 480 articles if one article cost $\$1.12\frac{1}{2}$?

$$\$1.12\tfrac{1}{2} = \$1\tfrac{1}{8} = \$\dfrac{9}{8} \quad \text{so} \quad \dfrac{\overset{60}{\cancel{480}}}{1} \times \$\dfrac{9}{\cancel{8}} = \$540$$

Alternate solution:

$$
\begin{array}{r}
480 \times \$1 = \$480 \\
+\; 480 \times \$\dfrac{1}{8} = \quad 60 \\
\hline
\$540
\end{array}
$$

ALIQUOT PARTS OF 100%

While any number can be used as a base for aliquot parts, the two most common bases are $1.00 and 100%. We saw in the previous discussion how aliquot parts of $1.00 can make calculations easier. Aliquot parts of 100% can also make many calculations easier. For example, what is $33\frac{1}{3}\%$ of 900? Working this out by using a decimal is very hard because the decimal equivalent of $33\frac{1}{3}\%$ is a repeating decimal (.333333...). So, the following calculation using an aliquot part is much better:

$$33\tfrac{1}{3}\% = \frac{1}{3} \quad \text{so} \quad \frac{900}{1} \times \frac{1}{3} = 300$$

CALCULATING ALIQUOT PARTS

If you do not recognize an aliquot part and do not have a table available, you can convert cents to aliquot parts of $1.00 (or percents to aliquot parts of 100%) by following the technique shown in the following examples. In effect, you are simply changing a decimal, or a percent, to a fraction.

Example 1
Convert $16\tfrac{2}{3}$ cents to an aliquot part of $1.00.

"16⅔ [or any #] is what fraction of a dollar?"

Solution

$$16\tfrac{2}{3}\text{¢} = \frac{3 \times 16 + 2}{3 \times 100} = \frac{50}{300} = \frac{1}{6}$$

Example 2
Convert $37\tfrac{1}{2}$ cents to an aliquot part of $1.00.

Solution

$$37\tfrac{1}{2}\text{¢} = \frac{2 \times 37 + 1}{2 \times 100} = \frac{75}{200} = \frac{3}{8}$$

Example 3
Convert $6\tfrac{1}{4}\%$ to an aliquot part of 100%.

Solution

$$6\tfrac{1}{4}\% = \frac{4 \times 6 + 1}{4 \times 100} = \frac{25}{400} = \frac{1}{16}$$

Self-Check
Convert the following to aliquot parts:

(a) $.08\tfrac{1}{3}$ (b) $87\tfrac{1}{2}\%$.

Solution

(a) $\$\dfrac{1}{12}$ (b) $\dfrac{7}{8}$

unit
5.4

Exercises

Find the cost of each of these purchases. Use aliquot parts of $1.00 to make each calculation and write the aliquot part in the parentheses next to each unit price. Round off answers to even cents where necessary.

1. 160 units @ $.25 (   ) = $_____

2. 300 units @ $.33$\frac{1}{3}$ (   ) = $_____

3. 252 units @ $.87$\frac{1}{2}$ (   ) = $_____

4. 900 units @ $.66$\frac{2}{3}$ (   ) = $_____

5. 540 units @ $.60 (   ) = $_____

6. 144 units @ $.08$\frac{1}{3}$ (   ) = $_____

7. 600 units @ $.16$\frac{2}{3}$ (   ) = $_____

8. 720 units @ $.87$\frac{1}{2}$ (   ) = $_____

9. 648 units @ $.12$\frac{1}{2}$ (   ) = $_____

10. 366 units @ $.83$\frac{1}{3}$ (   ) = $_____

11. 800 units @ $.80 (   ) = $_____

12. 336 units @ $.33$\frac{1}{3}$ (   ) = $_____

13. 672 units @ $.05 (   ) = $____

14. 331 units @ $.14$\frac{2}{7}$ (   ) = $____

15. 320 units @ $.40 (   ) = $____

Using aliquot parts of 100%, find the portion (remember that $B \times R = P$) in each of the following. Round answers to the nearest hundredth where necessary:

Base	Rate	Portion
16. 800	37$\frac{1}{2}$%	_____
17. 961	50%	_____
18. $600	33$\frac{1}{3}$%	$_____
19. $352	16$\frac{2}{3}$%	$_____
20. $720	12$\frac{1}{2}$%	$_____
21. $480	8$\frac{1}{3}$%	$_____
22. $301	66$\frac{2}{3}$%	$_____
23. $916	83$\frac{1}{3}$%	$_____
24. $200	5%	$_____
25. $618	75%	$_____

(handwritten) → What fraction of 100 % is The Rate?

Using aliquot parts, calculate the extension (units × unit price) for each of the following. Round final answers to even cents where necessary:

	Units	Unit Price	Extension
26.	1,200	$1.12\frac{1}{2}$	$_____
27.	800	$1.16\frac{2}{3}$	_____
28.	2,400	$2.08\frac{1}{3}$	_____
29.	4,500	$3.06\frac{2}{3}$	_____
30.	3,216	$1.06\frac{1}{4}$	_____
31.	900	$2.66\frac{2}{3}$	_____
32.	1,850	$41.37\frac{1}{2}$	_____
33.	3,624	$2.16\frac{2}{3}$	_____
34.	4,885	$1.20	_____
35.	1,800	$1.80	_____

Without referring to the chart in this unit, convert each of the following to aliquot parts of $1.00:

ANSWERS

36. $.80 =

37. $.33\frac{1}{3}$ =

38. $.20 =

39. $.87\frac{1}{2}$ =

40. $.66\frac{2}{3}$ =

41. $.37\frac{1}{2}$ =

42. $.06\frac{2}{3}$ =

43. $.83\frac{1}{3}$ =

44. $.98\frac{1}{3}$ =

45. $.16\frac{2}{3}$ =

36. _____
37. _____
38. _____
39. _____
40. _____
41. _____
42. _____
43. _____
44. _____
45. _____

Convert each of the following to aliquot parts of 100%:

46. $6\frac{1}{4}\% =$ **47.** $87\frac{1}{2}\% =$

48. $90\% =$ **49.** $33\frac{1}{3}\% =$

50. $40\% =$ **51.** $75\% =$

52. $37\frac{1}{2}\% =$ **53.** $12\frac{1}{2}\% =$

54. $5\% =$ **55.** $3\frac{1}{3}\% =$

46. _____

47. _____

48. _____

49. _____

50. _____

51. _____

52. _____

53. _____

54. _____

55. _____

Work Space

Name _____ **Date** _____

Applications

1. What is the total cost of 36 yards of fabric if one yard costs 1.37\frac{1}{2}$?

2. Todd McKing gets a commission of 6$\frac{1}{4}$% on everything he sells. Last month his total sales were $56,800. What is his commission for the month?

3. If a certain cleaning solution costs 2.66\frac{2}{3}$ a pound, what would be the cost of 9 pounds?

4. Kelli Reese paid $.87$\frac{1}{2}$ a quart for oil. How many quarts did she buy if her total bill came to $5.25?

5. John Lynn earns 6.12\frac{1}{2}$ an hour. How much does he earn in a week in which he worked 40 hours?

6. How much will 300 floor tiles cost at $.33$\frac{1}{3}$ for each tile?

7. If heating oil sells for 1.06\frac{2}{3}$ a gallon, how much will 15 gallons cost?

ANSWERS

1. _____
2. _____
3. _____
4. _____
5. _____
6. _____
7. _____

8. Of the 6,000 students at Maryvale College, $83\frac{1}{3}$% are from the United States. (a) How many students are from the United States? (b) How many students are from other countries?

9. Lakewood Community College has a total teaching faculty of 400 instructors. If $37\frac{1}{2}$% of the teaching faculty has a masters degree, how many instructors have masters degree?

10. Fill in this invoice:

ANSWERS

8. a. _____

   b. _____

9. _____

OCALA SUPPLY COMPANY
315 Central Avenue
Ocala, FL 32678

Sold to:
Florida Buildings Company
8000 College Parkway
Jacksonville, FL 32202

Item	Unit Cost	Extension
18 wrench sets	3.66\frac{2}{3}$	$ _____
24 lock sets	1.12$\frac{1}{2}$	_____
30 latch sets	4.50	_____
36 end holdings	.16$\frac{2}{3}$	_____
56 placeholders	.87$\frac{1}{2}$	_____
Total		$ _____

unit 5.5

Computing Markup and Selling Price

After completing Unit 5.5, you will be able to:

1. Compute markup based on the cost price.
2. Determine the rate of markup based on cost.
3. Compute markup based on the selling price.
4. Determine the rate of markup based on the selling price.

To start our discussion of markup and selling price, let's take an example.

Mark Price is thinking about driving to a nearby state to buy a used car. Mark plans to fix up the car and to resell it for a profit. Mark can buy the car for $800 and calculates that he will have to spend another $200 to get the car in condition to sell. Mark will also spend $10 for gas to get to the car and to have someone drive it back. Of course, Mark wants to make as much profit as he can on the transaction. But, he knows that he must set a price that is realistic. So, he first calculates his total investment in the car:

Cost . . . .$	800
Fixing up expenses	200
Gas	10
Total . . . .$	1,010

Mark now realizes that to make a profit, he must sell the car for more than his $1,010 investment in the car. Mark checks with several used car dealers and several banks to get the fair market value of a car like this. He finds that cars of this make and age are selling for around $1,600. Thus, Mark decided to set a selling price of $1,595.

In setting his selling price, Mark considered two factors: (1) how much he paid for the car (cost) and (2) how much he must add to the cost to cover expenses and to make a profit. This second factor is called **markup**—the amount added to the cost of goods to get the selling price.

Mark calculated the selling price of his car by adding a definite dollar amount to the cost of the car. His markup was the difference between his cost and the selling price. In business, markup is usually calculated by using a rate. There are two common bases used to determine markup: **markup based on cost**

Markup = expenses plus profit

and **markup based on the selling price.** We will look at both methods in the discussion that follows.

〜 MARKUP BASED ON COST

To calculate the amount of markup when cost is the base, multiply the cost of an item times the markup rate. Then you add the markup to the cost to get the selling price. In formula form, this looks like this:

	Cost of an item			Cost of an item
$\times$	markup rate		$+$	markup
$=$	markup		$=$	selling price

Example

On Monday morning, Sterling Clothing Store received a shipment of men's suits. Each suit had a cost of $90 and Sterling marks up all merchandise 40% of cost. What is the selling price of the suits?

Solution

Step 1: Calculate the markup:

$$
\begin{array}{ll}
\$90 & \text{cost} \\
\times\ .40 & \text{markup rate} \\
\hline
\$36 & \text{markup}
\end{array}
$$

Step 2: Add the markup, $36, to the cost to get the selling price:

$$
\begin{array}{ll}
\$90 & \text{cost} \\
+\ \ 36 & \text{markup} \\
\hline
\mathbf{\$126} & \textbf{selling price}
\end{array}
$$

Each business must decide what rate to use when marking up goods. Some merchandise, such as groceries, has a low markup because of high sales volume and heavy competition. Other merchandise, such as jewelry, usually has a much higher markup because the sales volume is usually low. Let's look at another example of markup based on cost.

Example

What is the selling price of an item with a cost of $600 and a markup rate of $33\frac{1}{3}\%$ based on cost?

Solution

$$
\begin{array}{ccccc}
\text{Cost} & \times & \text{markup rate} & = & \text{markup} \\
\$600 & \times & \frac{1}{3} & = & \$200
\end{array}
$$

> Remember: The aliquot part of $33\frac{1}{3}\%$ is $\frac{1}{3}$.

$$
\begin{array}{ccccc}
\text{Cost} & + & \text{markup} & = & \text{selling price} \\
\$600 & + & \$200 & = & \mathbf{\$800}
\end{array}
$$

FINDING THE RATE OF MARKUP BASED ON COST

In some problems, you will be given the selling price and the cost price and asked to find the rate of markup. To find the rate of markup when cost is the base, divide the amount of markup by the cost price. We can express this in formula form:

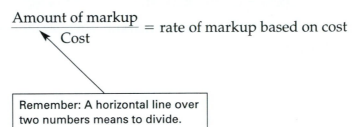

$$\frac{\text{Amount of markup}}{\text{Cost}} = \text{rate of markup based on cost}$$

Remember: A horizontal line over two numbers means to divide.

Example

Goods costing $500 were marked to sell for $600. What is the rate of markup based on cost?

Solution

Step 1: Determine the amount of markup by subtracting the cost from the selling price:

$$
\begin{array}{rl}
\$600 & \text{selling price} \\
-\ \underline{500} & \text{cost} \\
\$100 & \text{amount of markup}
\end{array}
$$

Step 2: Divide the amount of markup by the cost:

$$\frac{\$100}{\$500} = .2 \text{ or } 20\%$$

$$
\begin{array}{lrl}
\text{Proof:} & \$500 & \text{cost} \\
& \times\ \underline{.20} & \text{markup rate} \\
& \$100 & \text{amount of markup}
\end{array}
$$

MARKUP BASED ON SELLING PRICE

Markup can also be based on the selling price, or the retail price, of goods. Under this assumption, the selling price is calculated using this formula:

$$\frac{\text{Cost}}{100\% - \text{the markup rate}} = \text{selling price of goods}$$

Then, to get the amount of markup, subtract the cost of the goods from the selling price.

Example 1

An item with a cost of $600 is marked up 40% of the selling price. Determine the selling price.

Solution

$$\frac{\text{Cost}}{100\% - \text{the markup rate}} = \frac{\$600}{1.00 - .40} = \frac{\$600}{.60} = \mathbf{\$1,000}$$

Remember: Change your percents to decimals.

Proof: $1,000 × .40 = $400 markup
$1,000 − $400 = $600 cost

Example 2

An item with a cost of $420 is marked up 30% of the selling price. Determine the selling price and the amount of markup.

Solution

Determine the selling price:

$$\frac{\$420}{1.00 - .30} = \frac{\$420}{.70} = \mathbf{\$600}$$

Now subtract the cost from the selling price to get the markup:

$$\begin{array}{rl} \$600 & \text{selling price} \\ - \ \underline{420} & \text{cost} \\ \$180 & \textbf{markup} \end{array}$$

Proof: $\$600 \times .30 = \$180$ markup
$\$600 - \$180 = \$420$ cost

Self-Check

An item with a cost of $600 is marked up 40% of the selling price. Find (a) the selling price and (b) the amount of markup.

Solution

(a) $1,000   (b) $400

FINDING THE RATE OF MARKUP BASED ON SELLING PRICE

To find the rate of markup based on selling price, divide the amount of markup by the selling price. We can state this in formula form, as follows:

$$\frac{\text{Amount of markup}}{\text{Selling price}} = \text{rate of markup based on selling price}$$

Example

Twin City Furniture Company sold a bed for $300 that cost $225. What was the rate of markup based on the selling price?

Solution

$\$300 - \$225 = \$75$ markup
$\$75 \div \$300 = 25\%$ rate of markup based on selling price

Proof: $\$300 \times .25 = \$75$ markup

Self-Check

An item costing $200 was marked up to sell for $250. What was the rate of markup based on the selling price?

Solution

20%

Work Space

SECTION V   Percents

Name _____ Date_____

Exercises

Find the markup based on cost and the selling price for each of the following:

	Cost	Markup Rate on Cost	Amount of Markup	Selling Price
1.	$ 400	25%	$_____	$_____
2.	575	30%	_____	_____
3.	820	10%	_____	_____
4.	1,240	27%	_____	_____
5.	1,500	$33\frac{1}{3}\%$	_____	_____
6.	3,000	$37\frac{1}{2}\%$	_____	_____
7.	6,200	40%	_____	_____
8.	7,852	35%	_____	_____
9.	1,455	18%	_____	_____
10.	9,800	$12\frac{1}{2}\%$	_____	_____

Find the selling price and the markup based on selling price for each of the following. Round dollar amount to even cents where necessary:

Cost	Markup Rate of Selling Price	Selling Price	Amount of Markup
11. $ 800	20%	$_____	$_____
12. 1,200	40%	_____	_____
13. 784	30%	_____	_____
14. 228	25%	_____	_____
15. 600	$33\frac{1}{3}\%$	_____	_____
16. 2,480	15%	_____	_____
17. 4,400	22%	_____	_____
18. 3,200	60%	_____	_____
19. 2,700	10%	_____	_____
20. 6,400	36%	_____	_____

How do you deduce from, say, a 30% markup on the selling price, a 70% markup on the cost?

Cost = $500
Markup rate of selling price = 50%

$$\frac{500}{1-.50} = \frac{500}{.50} = \$1000$$

$$\frac{500+x}{.50} = \text{Selling price (where } x = \text{amt of markup)}$$

$$500/.50 + x/.50 = 1000 \ldots\ldots\ldots$$

For each of the following, find the amount of markup. Then find the rate of markup based on cost and the rate of markup based on the selling price. Round rates to the nearest tenth percent where necessary:

Selling Price	Cost	Amount of Markup	Rate of Markup Based on Cost	Rate of Markup Based on Selling Price
21. $ 800	$ 500	$ _300_	$\frac{300}{500} = \underline{60}$ %	$\frac{300}{800} = \underline{37.5}$ %

Proof: $\frac{500}{1-.375} = \frac{500}{.625} = 800$

Markup rate based on cost will always be greater than markup rate based on selling price (?)

22. 345	210	_____	_____	_____
23. 872	650	_____	_____	_____
24. 1,450	900	_____	_____	_____

Rate of Markup

	Selling Price	Cost	Amount of Markup	Rate of Markup Based on Cost	Rate of Markup Based on Selling Price
25.	3,000	2,200	_____	_____	_____
26.	4,200	3,500	_____	_____	_____
27.	8,000	6,000	_____	_____	_____
28.	4,500	3,000	_____	_____	_____

Applications

Name _____ Date _____

In the following application problems, round all rates to 3 places and all dollar amounts to even cents where necessary:

1. Goods costing $672 are marked up 25% of cost. What is the selling price?

2. Ben's Appliance Company bought a refrigerator for $890 and marked it up 30% of cost. What is (a) the amount of markup and (b) the selling price?

3. A store bought a couch for $600 and marked it up 40% of the selling price. What is (a) the selling price and (b) the markup?

4. Goods costing $6,500 were marked up to sell for $8,000. (a) What was the rate of markup based on cost? (b) What was the rate of markup based on the selling price?

1. _____

2. a. _____

   b. _____

3. _____

4. _____

5. The cost of merchandise in a small department store was $24,800. The selling price totaled $32,600.

   a. What was the markup?

   b. What was the rate of markup based on cost?

   c. What was the rate of markup based on the selling price?

6. A house was constructed at a cost of $60,000. It was then sold for $90,000. (a) What is the rate of gain based on cost? (b) What is the rate of gain based on the selling price?

7. A manufacturer sold goods to a wholesaler (a middleman) for $40,000. The wholesaler marked the goods up 20% of cost and sold them to a retailer. The retailer then marked up the goods 30% of the selling price and placed them on the shelf. (a) What was the cost of the goods to the retailer? (b) At what price did the retailer mark the goods for resell?

8. A sporting goods store bought baseball gloves for $144 a dozen. What is the price of each glove if the gloves were marked up 40% of the selling price?

ANSWERS

5. a. _____

   b. _____

   c. _____

6. a. _____

   b. _____

7. a. _____

   b. _____

8. _____

Computing Markdown and New Selling Price

After completing Unit 5.6, you will be able to:

1. Compute the amount of markdown.
2. Compute the rate of markdown.

Businesses often mark down the price of merchandise because of such things as promotional sales, seasonal changes, style changes, and inventory clearance. This reduction in the original selling price of merchandise is called **markdown,** and the selling price after the markdown is called the **new selling price:**

$$\text{Original selling price} - \text{markdown} = \text{new selling price}$$

Markdown can be expressed as a definite dollar amount. For example, a $500 item can be marked down by $50, which then gives a new selling price of $450 ($500 − $50). Or, markdown can be expressed as a rate (percent).

Example

An item with a selling price of $800 is marked down by 20%. What is the new selling price?

Solution

Original selling price $800	$800	
Markdown rate . . . .× .20	− 160	
Amount of markdown $160	**$640**	

new selling price

If a problem involves both a markup and a markdown, *the markup must be computed first.*

Example

What is the new selling price of a sofa costing $300, marked up 25% of cost, and then marked down by 40%?

Solution

$$
\begin{array}{ll}
\$300 & \text{cost} \\
+\ \ \underline{75} & \text{markup (\$300} \times .25) \\
\$375 & \text{original selling price} \\
-\ \underline{150} & \text{markdown (\$375} \times .40) \\
\$225 & \text{new selling price}
\end{array}
$$

Prices sometimes tend to move up and down. Any increase in price after a markdown is called a **markdown cancellation.** We can express this in formula form as follows:

Markdown − markdown cancellation = net markdown

Now we find the final selling price as follows:

Original selling price − net markdown = final selling price

Example

An item with an original selling price of $80 was marked down by 30% and later raised by 10%. What is the final selling price and the net markdown?

Solution

$$
\begin{array}{ll}
\$80.00 & \text{original selling price} \\
-\ \underline{24.00} & \text{markdown (\$80} \times .30) \\
\$56.00 & \\
+\ \underline{5.60} & \text{markdown cancellation (\$56} \times 10\%) \\
\mathbf{\$61.60} & \textbf{final selling price}
\end{array}
$$

Net markdown = $24.00 − 5.60 = $18.40

Note: You can also find the final selling price ($61.60) by subtracting the net markdown ($18.40) from the original selling price.

Self-Check

Goods with an original selling price of $3,400 were marked down 20%. What is (a) the amount of markdown and (b) the new selling price?

Solution

(a) $680   (b) $2,720

FINDING THE RATE OF MARKDOWN

Sometimes you will be given the original selling price and the new selling price and asked to calculate the rate (percent) of markdown. To do this, you divide the amount of markdown by the original selling price. In formula form, this looks like this:

$$\frac{\text{Markdown}}{\text{Original selling price}} = \text{rate of markdown}$$

Example

Goods with an original selling price of $6,000 were marked down to sell for $4,500. What was the rate of markdown?

Solution

Step 1: Find the amount of markdown:

$6,000   original selling price
− 4,500   new selling price
$1,500   markdown

Step 2: Divide the markdown by the original selling price:

$$\frac{\$1,500}{\$6,000} = .25 = 25\%$$

Proof: $6,000 × .25 = $1,500

Self-Check

Goods with an original selling price of $5,000 were marked down to sell for $4,500. What is the rate of markdown?

Solution

10%

Work Space

Exercises

For each of the following, compute the markdown and the new selling price.
Round all dollar amounts to even cents where necessary:

	Original Selling Price	Rate of Markdown	Markdown	New Selling Price
1.	$ 800	25%	$_____	$_____
2.	950	30%	_____	_____
3.	780	10%	_____	_____
4.	1,200	$33\frac{1}{3}\%$	_____	_____
5.	4,000	15%	_____	_____
6.	6,450	40%	_____	_____
7.	8,455	10%	_____	_____
8.	8,000	$37\frac{1}{2}\%$	_____	_____
9.	12,800	32%	_____	_____
10.	15,600	45%	_____	_____
11.	45,200	28%	_____	_____
12.	14,250	36%	_____	_____

Find the rate of markdown for each of the following. Round rates to the nearest tenth percent where necessary.

	Original Selling Price	New Selling Price	Rate of Markdown
13.	$ 600	$ 450	_____%
14.	900	600	_____%
15.	1,400	1,200	_____%
16.	2,500	2,000	_____%
17.	4,250	2,750	_____%
18.	6,240	4,840	_____%
19.	5,000	4,000	_____%
20.	3,800	3,200	_____%

Work Space

Applications

1. An item with a cost of $600 was marked up 40% of the selling price. It was then marked down 20% and sold. What was (a) the original selling price, (b) the markdown, and (c) the new selling price?

2. Goods with an original selling price of $4,200 were marked down 25%. What is the new selling price?

3. An item with a cost of $500 was marked up 40% of cost. It was then marked down 15% and sold. Find (a) the markup, (b) the original selling price, (c) the markdown, and (d) the new selling price.

4. Goods with an original selling price of $2,000 were marked down to sell for $1,800. What was the rate of markdown?

5. If a 20% markdown amounts to $27, what was the original selling price? (*Hint: P = $27, R = 20%, find B.*)

6. The new selling price of an item is $48. What was the rate of markdown if the original selling price was $60?

ANSWERS

1. a. _____
   b. _____
   c. _____
2. _____
3. a. _____
   b. _____
   c. _____
   d. _____
4. _____
5. _____
6. _____

7. A bicycle is marked down $40, which is 40% of the original selling price. (a) What was the original selling price? (b) What is the new selling price?

7. a. _____

   b. _____

8. _____

8. Harry's Clothing Store marked down shoes from $60 to $40. What was the rate of markdown?

unit 5.7

Ratios and Proportions

After completing Unit 5.7, you will be able to:

1. Express numbers as ratios.
2. Determine if two ratios are proportional.

 ## RATIOS

A **ratio** is a comparison of one number to another number. You write a ratio either by separating the two numbers by the word *to*, separating the numbers by a colon (:), or expressing the relationship as a fraction. For example, if Coast State Community College has 60 full-time instructors and 10 part-time instructors, the ratio of full-time to part-time instructors is

60 to 10   or   60:10   or   60/10

The expression *60/10* can be reduced to 6/1 and is commonly expressed as 6:1, which means that there are six full-time instructors for each one part-time instructor. Likewise, a ratio of 3:1 means that the first number is three times as large as the second number. That is, if the second number is 100, then the first number will be 300.

Ratios are used extensively in business to express various relationships, such as the ratio of female workers to male workers, the ratio of older workers to younger workers, the ratio of retired people to working people, and the ratio of a company's assets to its debts.

PROPORTIONS

A **proportion** is the comparison of two ratios. For example, in our example of the number of full-time instructors compared to part-time instructors, we expressed the relationship as 60:10, which we reduced to 6:1. When compared, these two ratios form a proportion, as follows:

60/10 = 6:1

We read this expression as *60 is to 10 as 6 is to 1.* Likewise, the ratio of 12:60 can be reduced to 1:5 and forms the proportion: 12/60 = 1/5, which we read as *12 is to 60 as 1 is to 5.* Since the relationship is the same between the two sets of numbers, their cross products are equal. For example,

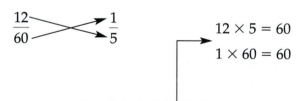

$12 \times 5 = 60$

$1 \times 60 = 60$

In cross multiplication, the numerator of each ratio is multiplied by the denominator of the other ratio.

Name _____ Date_____

Exercises

Express the following as ratios:

1. Ratio of Americans to Europeans if 30 Americans participate with 270 Europeans in an outdoor concert.

2. Ratio of sugar to flour if a recipe calls for one pound of sugar and 8 pounds of flour.

3. Ratio of men to women if an aerobics class of 45 has 5 men.

4. Ratio of women to men in the aerobics class in problem 3.

5. Ratio of retired persons to working persons in Florida if there are 2,000,000 retired persons and 7,000,000 working persons.

Determine if the following ratios are proportional:

6. 3/9, 12/27

7. 7/8, 14/15

8. 5/16, 7/15

9. 30/50, 270/450

10. 1:3, 50:150

11. 4:5, 16:20

12. 8:9, 24:36

Name _____ Date_____

Applications

1. Spencer Investments Company has 600 total employees, broken down as follows: 250 clerks, 300 brokers, 5 managers, and 45 assistant managers. What is the ratio of:

   a. clerks to total employees?

   b. clerks to brokers?

   c. managers to total employees?

   d. managers to assistant managers?

   e. assistant managers to managers?

   f. managers to brokers?

   g. brokers to total employees?

2. The CEO of Dysan earned $800,000 last year. The average employee earned $25,000. How much did the CEO earn for each dollar earned by the average employee.

3. The year before free agency was started in professional baseball, the average annual player's earnings was $51,000. Fifteen years later the average annual salary was $1,100,000. What is the ratio of earnings before free agency to the earnings after free agency?

ANSWERS

1. a. _____

   b. _____

   c. _____

   d. _____

   e. _____

   f. _____

   g. _____

2. _____

3. _____

343

4. Smith Foundries has 75 production workers and 15 office workers. What is (a) the ratio of office workers to production workers and (b) the ratio of production workers to office workers?

5. If Crown Burger can produce 125 hamburger patties from 30 pounds of ground beef, how many patties can be produced from 50 pounds of ground beef?

Work Space

section

VI

Math for Consumer and Business Use

unit 6.1

Bank Checking Accounts

After completing Unit 6.1, you will be able to:

1. Write checks.
2. Prepare check stubs.

A bank is a financial institution that offers various services to its customers. One of the most common and most important services offered by banks is the **checking account.** A checking account is an amount of cash that a person or business has **on deposit** with a bank. The use of bank checking accounts, for personal use as well as business use, has become a near universal business practice. Not only does a checking account provide convenience in making cash payments, it provides control and protection of cash.

〰 BANK CHECKS

A **check** is a written order directing a bank to pay a specified sum of money to a specified person or business. A check is said to be **drawn** against the account of the person writing the check. Thus, a person or business who writes a check is called the **drawer.** The person or business to whom the check is made payable is called the **payee.** And the bank honoring the check for payment is called the **drawee.**

Checks vary in form, depending on the needs of the user. A common form of check has a stub attached for recording the date, amount, and purpose of the check. This type of check looks like the following one.

CHECK WRITING PROCEDURE

Errors that can be costly in terms of time and money can be minimized—or avoided completely—by using established procedures when writing checks. For best results, these procedures should be followed:

1. The check stub should be filled out first; otherwise this important form could be overlooked and a check could be mailed without a record.
2. The same date entered on the check stub should be entered on the **Date** line of the check.
3. The name of the payee should be clearly written on the **Pay to the Order Of** line.
4. The amount of the check should be written numerically after the printed dollar sign (on the same line with the name of the payee). The numbers should be written as close as possible to the dollar sign to eliminate the possibility of alteration.
5. The amount of the check should be written out in words on the line directly under the **Pay to the Order Of** line. Dollars are written out in words, and cents are written in terms of hundredths of a dollar. A line should be drawn to fill in empty space.
6. The check should be signed by an authorized person.
7. Corrections should not be made on a check. Should a mistake be made, write the word "VOID" on both the check and the check stub and write a new check.

To illustrate the check writing process, let's look at the following example, which shows several checks written by Thomas B. Anderson during June 20XX.

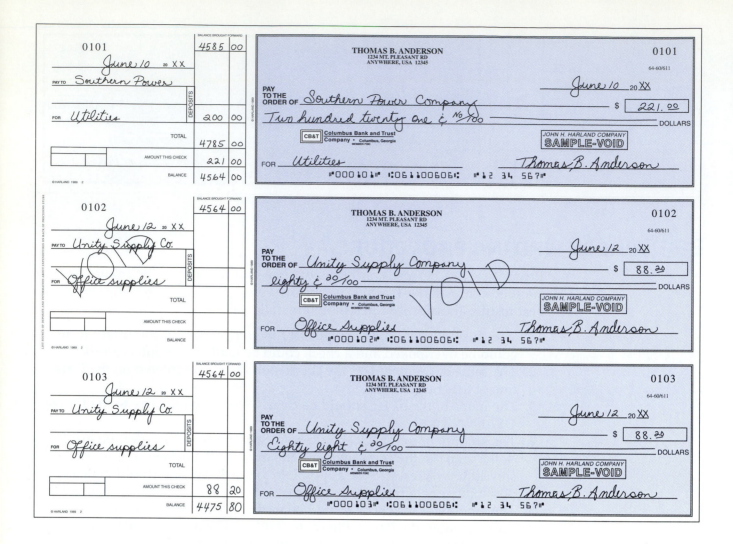

MAKING BANK DEPOSITS

To minimize the amount of cash on hand, and thus to minimize the risk of loss, most businesses make daily deposits of cash and checks received. A **deposit slip** or **deposit ticket** should be prepared each time a deposit is made. A completed deposit slip follows. Notice that space is provided for listing currency (paper money), coins, and checks that are being deposited.

JAN MARCANO, INTERIOR DECORATOR		
1901 Ried Drive		
Columbus, GA 31904		

64-60 / 601

Currency	153	00
Coin	2	35
Checks		
3—448	32	50
2—112	12	15
Total	200	00

Date June 10 20 XX

CB&T Columbus Bank and Trust
Company • Columbus, Georgia
MEMBER FDIC

⑆060100666⑆ ⑈12 17 860⑈

Self-Check

Write a check for $250 payable to Binks Supply Company. Complete the check stub assuming the balance before the payment was $3,275. The date is July 20.

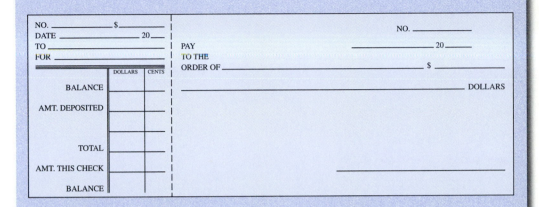

Solution

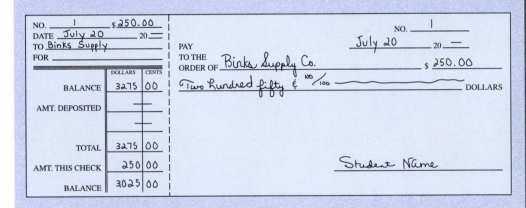

Name _____ Date_____

Exercises

1. Fill in the balances in the following problem that shows the deposits and checks that were recorded on a series of check stubs:

Deposit	6,200.00
Check 1	401.98
Balance	
Check 2	304.88
Balance	
Check 3	677.11
Balance	
Deposit	975.80
Balance	
Check 4	1,245.66
Balance	
Check 5	433.55
Balance	
Deposit	3,458.95
Balance	
Check 6	567.89
Balance	

2. The check stubs shown here are from Bi-City Paint Company. The amount of each check and deposit has been recorded. Complete the stubs.

NO. 102		$ 10.50
DATE January 2		20—
TO Kim Supply		
FOR Supplies		
	DOLLARS	CENTS
BALANCE	691	02
AMT. DEPOSITED	—	
TOTAL		
AMT. THIS CHECK	10	50
BALANCE		

NO. 105		$ 67.80
DATE January 9		20—
TO Dale Powers		
FOR Lawn Care		
	DOLLARS	CENTS
BALANCE		
AMT. DEPOSITED		
TOTAL		
AMT. THIS CHECK	67	80
BALANCE		

NO. 108		$ 225.00
DATE January 19		20—
TO Kathy Flose		
FOR Salary		
	DOLLARS	CENTS
BALANCE		
AMT. DEPOSITED	800	00
TOTAL		
AMT. THIS CHECK	225	00
BALANCE		

NO. 103		$ 49.50
DATE January 3		20—
TO Postal Service		
FOR Stamps		
	DOLLARS	CENTS
BALANCE		
AMT. DEPOSITED		
TOTAL		
AMT. THIS CHECK	49	50
BALANCE		

NO. 106		$_____
DATE _____		20—
TO _____		
FOR _____		
	DOLLARS	CENTS
BALANCE		
AMT. DEPOSITED	VOID	
TOTAL		
AMT. THIS CHECK		
BALANCE		

NO. 109		$ 195.00
DATE January 21		20—
TO Head Service Co.		
FOR Repairs		
	DOLLARS	CENTS
BALANCE		
AMT. DEPOSITED		
TOTAL		
AMT. THIS CHECK	195	00
BALANCE		

NO. 104		$ 418.00
DATE January 6		20—
TO Central Power		
FOR Power Bill		
	DOLLARS	CENTS
BALANCE		
AMT. DEPOSITED	841	00
TOTAL		
AMT. THIS CHECK	418	00
BALANCE		

NO. 107		$ 80.00
DATE January 15		20—
TO Kim Supply		
FOR Supplies		
	DOLLARS	CENTS
BALANCE		
AMT. DEPOSITED	2140	00
TOTAL		
AMT. THIS CHECK	80	00
BALANCE		

NO. 110		$ 225.00
DATE January 30		20—
TO Kathy Flove		
FOR Salary		
	DOLLARS	CENTS
BALANCE		
AMT. DEPOSITED		
TOTAL		
AMT. THIS CHECK	225	00
BALANCE		

unit 6.1

Applications

1. Using the check shown here, answer the following questions:

    a. Who is the payee?

    b. Who is the drawer?

    c. Who is the drawee?

    d. What is the amount of the check?

ANSWERS

1. a. _____

   b. _____

   c. _____

   d. _____

Susan Holloway
1901 Two Lakes Way
Tucson, AZ

No. 562

00-00
611

Date _July 6_ 20 XX

PAY TO THE ORDER OF _Beth Dutoit_ $ | 20.00 |

Twenty & 00/100 _____ DOLLARS

TUCSON NATIONAL BANK

Ben Brevard

⑆000562⑆ ⑈0611006161⑈ ⑆12 12 860⑈

---

2. The deposits and cash expenditures of Case Equipment Company for June 20XX follow. Using the checkbook provided, write a check for each expenditure. Be sure to fill in the check stub before you write the check:

June   1:   balanced forward, $6210.58.

         2:   wrote check #105 for $1,250 to Carter Realty for June rent.

         8:   made a deposit of $800.

         8:   wrote check #106 to Lot Service Co. for repairs, $238.50.

       12:   made a deposit of $2,680.

       12:   wrote check #107 to cash for salaries, $2,415.

       18:   made a deposit of $6,950.

       18:   wrote check #108 to Central Power Co., for utilities, $1,200.

       26:   wrote check #109 to cash for salaries, $2,415.

| NO. _____ $_____ |
| DATE _____ 20___ |
| TO _____ |
| FOR _____ |

	DOLLARS	CENTS
BALANCE		
AMT. DEPOSITED		
TOTAL		
AMT. THIS CHECK		
BALANCE		

NO. _____

_____ 20 _____

PAY
TO THE
ORDER OF _____ $ _____

_____ DOLLARS

_____

| NO. _____ $_____ |
| DATE _____ 20___ |
| TO _____ |
| FOR _____ |

	DOLLARS	CENTS
BALANCE		
AMT. DEPOSITED		
TOTAL		
AMT. THIS CHECK		
BALANCE		

NO. _____

_____ 20 _____

PAY
TO THE
ORDER OF _____ $ _____

_____ DOLLARS

_____

| NO. _____ $_____ |
| DATE _____ 20___ |
| TO _____ |
| FOR _____ |

	DOLLARS	CENTS
BALANCE		
AMT. DEPOSITED		
TOTAL		
AMT. THIS CHECK		
BALANCE		

NO. _____

_____ 20 _____

PAY
TO THE
ORDER OF _____ $ _____

_____ DOLLARS

_____

NO. _____ $_____
DATE _____ 20 ___
TO _____
FOR _____

	DOLLARS	CENTS
BALANCE		
AMT. DEPOSITED		
TOTAL		
AMT. THIS CHECK		
BALANCE		

NO. _____

_____ 20 _____

PAY
TO THE
ORDER OF _____ $ _____

_____ DOLLARS

_____

NO. _____ $_____
DATE _____ 20 ___
TO _____
FOR _____

	DOLLARS	CENTS
BALANCE		
AMT. DEPOSITED		
TOTAL		
AMT. THIS CHECK		
BALANCE		

NO. _____

_____ 20 _____

PAY
TO THE
ORDER OF _____ $ _____

_____ DOLLARS

_____

NO. _____ $_____
DATE _____ 20 ___
TO _____
FOR _____

	DOLLARS	CENTS
BALANCE		
AMT. DEPOSITED		
TOTAL		
AMT. THIS CHECK		
BALANCE		

NO. _____

_____ 20 _____

PAY
TO THE
ORDER OF _____ $ _____

_____ DOLLARS

_____

Work Space

SECTION VI   Math for Consumer and Business Use

unit 6.2

Bank Reconciliation

After completing Unit 6.2, you will be able to:

1. Prepare a bank reconciliation statement.

 **THE BANK STATEMENT**

A **bank statement** is a report that gives the status of a depositor's account according to the bank's records. The bank statement is prepared periodically, usually once a month. The statement lists the amounts deposited and withdrawn, interest earned on the account, if any, charges made by the bank against the account, and the balance of the account as of the date of the statement. Included with the bank statement will be all **canceled checks,** that is, checks that have cleared the account for payment during the period covered by the statement.

The following bank statement was received by Jan Marcano on July 3, 20XX.

Bank Statement

To: Jan Marcano
1901 Ried Drive
Columbus, GA 31904

Columbus Bank & Trust
Columbus, GA

Acct. No. 12 17 860

June 30, 20XX

Checks	Deposits	Date	Balance-$4,200
$200.00		6-2-XX	$5,120.00
125.00		6-4-XX	4,995.00
301.00		6-7-XX	4,694.00
109.00		6-9-XX	4,585.00
221.00	$ 200.00	6-10-XX	4,564.00
68.20		6-12-XX	4,475.80
322.80		6-18-XX	4,153.00
	1,400.00	6-22-XX	5,553.00
920.00		6-25-XX	4,633.00
856.10		6-28-XX	3,776.90
18.00 SC			
15.00 IP			3,743.90
	456.10	6-28-XX	4,200.00

Legend
DM = debit memorandum   SC = service charge   NSF = nonsufficient funds
CM = credit memorandum   IP = imprinting checks

RECONCILING THE BANK STATEMENT

The bank statement, as stated above, is a statement of the depositor's account according to the bank's records. And the checkbook balance is the depositor's record of the account. Thus, it would seem that the two records should agree in amount. But, in practice, this is rarely the case. This is not because of mistakes or inattention on the part of the depositor or the bank but, instead, is a result of several factors. Let's look at the common causes of disagreement between the bank statement balance and the checkbook balance.

1. **Deposits in transit.** Deposits in transit, also called **late deposits,** are deposits that did not reach the bank's accounting department in time to be added to the current bank statement. They appear in the checkbook, but not on the bank statement.

2. **Outstanding checks.** When a check is written by the depositor, the amount of the check should immediately be entered in the checkbook as a reduction in the checkbook balance. Once written, however, a check could take several days before it reaches the bank for payment. If the bank statement is prepared before a check clears the account for payment, the check will not show up on the bank statement. The check, then, is said to be outstanding.

3. **Service charges.** Banks usually charge a fee for providing checking accounts. This fee, called a service charge, is subtracted directly from the depositor's balance and is entered on the bank statement.

4. **Other bank charges.** There are certain other charges that banks make on a regular basis. These include charges for printing the checks, and charges for insufficient funds checks (ISF), that is, checks written for an amount greater than the balance of the account.

5. **Errors.** Errors are a common cause of disagreement between the checkbook balance and the bank statement balance. Most are made by the depositor, because banks today use electronic equipment for processing bank transactions.

When the bank statement balance and the checkbook balance do not agree, they must be reconciled, that is, brought into agreement. To reconcile the statement, follow these steps:

1. Add the amount of deposits in transit to the bank statement.
2. Subtract the amount of outstanding checks from the bank statement balance.
3. Subtract all charges appearing on the bank statement from the checkbook balance.

To illustrate the bank reconciliation process, let's look again at Jan Marcano's bank statement on page 358. According to the bank statement, Jan's account has a balance of $4,200. But, on the same date, Jan's checkbook shows a balance of $3,831. The two records were brought into agreement as follows:

1. By comparing each deposit recorded in her checkbook with the deposits appearing on the bank statement, Jan found that a deposit of $600 made on July 2 had not reached the bank's accounting department in time to be entered on the statement. So, this deposit is outstanding.
2. By arranging her canceled checks in numeric order and comparing each check recorded in the checkbook with those appearing on the bank statement, Jan found that several checks had not cleared her account. Specifically, the following checks were outstanding:

Check No.	Amount
221	$ 75
222	125
225	102
226	700

3. Jan checked the bank statement for any charge that may have been made against her account. She discovered that the bank had charged her account $18 for a service charge and $15 for printing checks.

Based on this analysis, Jan prepared the reconciliation statement shown below.

JAN MARCANO, INTERIOR DECORATOR

Bank Reconciliation
July 3, 20XX

Balance per bank statement	$4,200
Add deposit in transit	600
	$4,800
Deduct outstanding checks:	
#221 . . . .$ 75	
#222 125	
#225 102	
#226 700	1,002
Adjusted bank balance	$3,798
Balance per checkbook	$3,831
Deduct:	
Service charge . . . .$ 18	
Imprinting checks 15	33
Adjusted checkbook balance	$3,798

Bank Reconciliation Statement

Self-Check

Stephanie Goodman received her bank statement on July 31, 20XX. According to her statement, Stephanie has a balance of $3,600 in her checking account. According to Stephanie's checkbook, however, she has a balance of $3,687. By closely observing her records, Stephanie discovered the following:

(a) A $400 deposit did not reach the bank's bookkeeping department in time to be entered in her account when the bank statement was prepared.

(b) The following checks were outstanding:
    #112 for $125
    #114 for   50
    #115 for   40
    #116 for   110

(c) A service charge of $12 was made by the bank against Stephanie's account.

Prepare a bank reconciliation in good form.

Solution

STEPHANIE GOODMAN

Bank Reconciliation
July 31, 20XX

Balance per bank statement		$3,600
Add deposit in transit		400
		$4,000
Deduct outstanding checks:		
#112	$125	
#114	50	
#115	40	
#116	110	325
Adjusted bank balance		$3,675
Balance per checkbook		$3,687
Deduct service charge		12
Adjusted checkbook balance		$3,675

unit 6.2

Exercises

1. When reconciling a bank statement, state whether each of the following should be (a) added to the bank statement balance, (b) subtracted from the bank statement balance, (c) added to the checkbook balance, or (d) subtracted from the checkbook balance.

   a. _____ outstanding checks

   b. _____ deposits in transit

   c. _____ service charge

   d. _____ fee for printing checks

   e. _____ ISF check charge

2. List several factors that may prevent the bank statement balance from agreeing with the checkbook balance:

3. List, in order, the steps involved in reconciling a bank statement:

Work Space

Name _____ **Date** _____

Applications

1. William Reynolds received his bank statement on June 30, 20XX. According to his statement, William has a balance of $2,400 in his checking account. According to William's checkbook, however, his balance is $2,788. Upon closer observation, William discovered the following:

   a. An $800 deposit had not reached the bank in time to be entered in William's account when the statement was prepared.

   b. The following checks were outstanding:

   #144 for $200
   #145 for   25
   #147 for   180
   #148 for   15

   c. A service charge of $8 was made against William's account by the bank.

   Prepare a bank reconciliation statement <mark>in good form.</mark>

2. Reconcile the bank statements in the following problems:

a. Balance shown on the bank statement of Jill House, $655.45; balance as shown in the checkbook, $518.90.

Checks outstanding:

#135	. . . .$18.50
#138	. . . . .22.80
#142	. . . . .51.40
#145	. . . . .19.02
#147	. . . . .32.48

Service charge, $7.65.

b. Balance shown on the bank statement of Tri-City Paint Company, March 31, 20XX, $4,280.88; balance as shown in the checkbook, $4,035.13.

Deposit in transit, $312.00.

Outstanding checks:

#244	. . . .$125.00
#245	. . . . 75.60
#248	. . . . 34.89
#250	. . . . .325.00
#253	. . . . 19.21

Charge for printing checks, $12.50.

Service charge, $9.45.

unit 6.3

Computing Gross Pay

After completing Unit 6.3, you will be able to:

1. Compute gross pay for employees.

The total amount an employee earns during a payroll period is called **gross pay** or **gross earnings.** Gross pay can be determined in a variety of ways. In this unit, we look at the two most common methods of determining gross pay: (1) an **hourly wage** (or **hourly rate**) and (2) a fixed **salary.** Both methods are discussed next.

≈≈≈ HOURLY WAGES

Most employees who work for an hourly wage are covered by the **Fair Labor Standards Act,** which is a federal law that sets the minimum hourly wage and the standard workweek. At this writing, the *minimum* hourly wage is $5.15 for a standard workweek of 40 hours. If an employee works more than 40 hours in a given workweek, the employee must be paid at least one and one-half times the regular hourly rate for all hours worked in excess of 40. This is called **overtime.** We will look at overtime shortly. For the moment, let's look at an example of finding gross pay when overtime does not apply.

Example
Joy Bodiford earns an hourly wage of $5.50. Last week Joy worked 40 hours. What was her gross pay?

Solution

$$\frac{\text{Hours worked} \times \text{rate per hour} = \text{gross pay}}{40 \quad\quad \times \quad\quad \$5.50 \quad = \quad \$220.00}$$

In this example, Joy earns $5.50 an hour, which is more than the minimum wage. Remember that the key word here is *minimum.* Many companies pay well in excess of the minimum wage as a matter of company policy. And some companies are not required to pay the minimum wage—generally, these are companies that are not involved in interstate commerce.

Let's now look at an example where an employee works more than 40 hours in a week. Remember that the law requires the employer to pay at least one and one-half times the regular hourly rate for overtime. Thus, the overtime rate can be found by multiplying the regular hourly rate by 1.5.

Example

Clayton Morrow earns an hourly wage of $6.00. Last week he worked a total of 44 hours. What is his gross pay?

Solution

Step 1: Find his regular pay for 40 hours:

$$\underline{\text{Regular hours}} \times \underline{\text{hourly wage}} = \underline{\text{regular pay}}$$
$$40 \quad\quad \times \quad \$6.00 \quad = \quad \$240$$

Step 2: Find his overtime pay:

$$\underline{\text{Hours over 40}} \times \underline{\text{overtime rate}} = \underline{\text{overtime pay}}$$
$$4 \quad\quad \times \quad \$9\ (\$6 \times 1.5) = \quad \$36$$

Step 3: Add regular pay and overtime pay to get his gross pay:

$$\$240 + \$36 = \mathbf{\$276}$$

Since overtime pay is at least one and one-half times the regular rate of pay, it is often called **time and a half.**

Employers sometimes calculate overtime at a rate other than 1.5 times the regular hourly rate. For example, some companies pay overtime if total hours worked in a day exceed 8, even though total hours for the week do not exceed 40. Additionally, some companies pay double time (and sometimes triple time) for work on weekends and holidays. But, it is important to remember that such plans are a matter of individual company policy, not the law. Let's look at an example of an employee who earns double time for any work on Saturday.

Example

Reggie South earns a regular hourly wage of $8.00. Reggie gets overtime for any hours worked over 40, and double time for any hours worked on Saturday. Last week Reggie worked 48 hours, 4 of which were on Saturday. What is his gross pay?

Solution

Step 1: Find his regular pay for 40 hours:

$$40 \times \$8 = \$320$$

Step 2: Find his overtime pay (4 regular overtime hours):

$$4 \times \$12\ (\$8 \times 1.5) = \$48$$

Step 3: Find his double time pay (4 hours on Saturday):

$$4 \times \$16\ (\$8 \times 2) = \$64$$

SECTION VI   Math for Consumer and Business Use

Step 4: Add the three amounts to find his gross pay:

$$\$320 + \$48 + \$64 = \mathbf{\$432}$$

 **SALARIES**

Salary is usually stated as a certain amount of money paid to an employee on an annual basis. Salaried employees are usually paid the same, whether hours worked are 40 and under, or over 40. You find the amount of a salaried person's paycheck by dividing the annual salary by the number of pay periods in the year. Pay periods for salaried people are usually on a **weekly, biweekly, semimonthly,** or **monthly** basis. To illustrate, let's say that an employee is hired at an annual salary of $16,900. We can calculate gross pay per period as follows:

Pay Period	Number of Pay Periods in a Year	Gross Pay per Pay Period
Weekly	52	$ 325.00 ($16,900 ÷ 52)
Biweekly	26	650.00 ($16,900 ÷ 26)
Semimonthly	24	704.17 ($16,900 ÷ 24)
Monthly	12	1,408.33 ($16,900 ÷ 12)

Self-Check

Dave Nolin earns a regular hourly wage of $9.00 an hour with time and a half for all hours over 40 in a week. Last week he worked 48 hours. What is his gross pay for the week?

Solution

$468

Work Space

Exercises

Name _____ Date _____

1. Calculate regular pay and gross pay for each of the following employees. All employees are paid time and a half if hours exceed 40:

Employee	Hourly Pay	Hours Worked	Regular Pay	Total Pay
A	$8.00	44	$_____	$_____
B	6.25	45	_____	_____
C	4.00	25	_____	_____
D	5.00	50	_____	_____
E	7.50	48	_____	_____
F	6.90	52	_____	_____
G	7.40	40	_____	_____

Employee	Hourly Pay	Hours Worked	Regular Pay	Total Pay
H	4.75	46	_____	_____
I	9.50	54	_____	_____
J	8.50	47	_____	_____

2. Find the gross pay of each of the following employees. All employees are paid time and a half for hours over 40 and double time for Sunday work:

Employee	Hours Worked M T W T F S S	Total Hours	Hourly Rate	Regular Pay	Overtime Pay	Gross Pay
A. Smith	8 8 7 8 8 7	_____	$ 6.50	$_____	$_____	$_____
B. Kaye	8 8 8 8 8 4	_____	8.25	_____	_____	_____
S. Snead	7 6 9 9 9 2	_____	6.00	_____	_____	_____
T. Lee	8 8 8 8 8	_____	12.00	_____	_____	_____
H. Dobbs	8 8 8 8 8 4 6	_____	8.00	_____	_____	_____
K. Frost	8 8 8 9 8	_____	10.00	_____	_____	_____

Applications

1. James Sanders makes $8.25 an hour and his employer is covered by the Fair Labor Standards Act. Last week James worked a total of 49 hours. What is his gross pay?

ANSWERS

1. _____

2. _____

3. _____

4. _____

2. Sue Chappell worked 34 hours in one week. What is her gross pay if her hourly wage is $5.50?

5. _____

3. Last week Lin Todd worked 40 hours at an hourly wage of $8.00. What is his gross pay?

4. Kathy James is paid 1.5 times her regular pay for all hours worked over 40 in a week. Last week Kathy worked 45 hours and is paid a regular hourly wage of $9.00. What was her gross pay for the week?

5. Patricia Belser has a salaried job that pays her $450 a week. Last week Patricia worked 42 hours. What is her gross pay?

6. Rob Abrams was just hired at an annual salary of $18,720. What is Rob's gross pay if his pay period is (a) weekly, (b) biweekly, (c) semimonthly, and (d) monthly.

ANSWERS

6. a. _____

b. _____

c. _____

d. _____

7. _____

8. _____

7. Tammy Lewis gets paid 1.5 times her regular hourly rate if she works more than 8 hours in a day, even though she may not work more than 40 hours in the week. Last week she worked 8 hours on Monday, 10 hours on Tuesday, 10 hours on Wednesday, and 6 hours on Thursday. What is her gross pay if her regular hourly wage is $6.50?

8. Sean Wilson earns $9.00 an hour and gets time and a half for hours over 40 and double time for work on holidays. Last week he worked 54 hours, 8 of which were on a holiday. What is his gross pay?

Work Space

unit 6.4

Payroll Deductions and the Payroll Register

After completing Unit 6.4, you will be able to:

1. Compute payroll deductions.
2. Compute net pay.
3. Prepare a payroll register.

In Unit 6.3, we learned how to calculate gross pay for employees who work for a salary and employees who work for an hourly wage. In this unit, we are going to study how to find net pay. Net pay (also called take-home pay) is the amount left after all payroll deductions have been made; it is the actual amount of an employee's paycheck.

Payroll deductions can be grouped into two categories: (1) those required by law and (2) those agreed to by the employee and the employer. Let's look at those required by law first.

PAYROLL DEDUCTIONS REQUIRED BY LAW

The federal government and most state governments require employers to withhold income taxes from the earnings of employees. The federal government also requires employers to withhold FICA (social security) from the wages of employees.

Federal Income Tax

The main source of revenue for our country comes from the income tax imposed on individual employees. The amount of federal income tax an employee pays depends on three factors: (1) the employee's gross pay, (2) the employee's marital status, and (3) the number of **withholding allowances** claimed by the employee.

We calculated gross pay in Unit 6.3. An employee's marital status, for the purpose of withholding income tax, is either married or single. A withholding allowance, also called an **exemption,** is an amount of earnings that is not subject to income taxes. An exemption is allowed for the employee and each person for whom the employee provides support.

To get the actual amount of federal income tax to be withheld, most employers use tax tables published by the Internal Revenue Service (an agency of the federal government). Tax tables are part of a publication called **Circular E,** which provides tax tables for weekly, biweekly, semimonthly, monthly, and daily or miscellaneous payroll periods for married and single persons. Tax tables for *Single Persons—Weekly Payroll Period* and *Married Persons—Weekly Payroll Period* are shown on pages 377 and 378.

To illustrate the use of tax tables, we will use Lang Kirkland as our example. Lang is married and claims three withholding allowances (one for himself, one for his wife, and one for his child). Lang works by the week and last week his gross pay was $350. To figure Lang's federal income tax, look at the Married Persons—Weekly Payroll table shown on page 378. Lang's pay, $350, is in the bracket of "at least $350 but less than $360." So, with three withholding allowances, Lang's federal income tax is $28. Find this amount in the table.

State and Local Income Taxes

As stated earlier, most state governments (and also some city governments) require employers to withhold an income tax from the earnings of employees. The rates of these taxes vary so greatly from state to state that it would not be possible to show all the various tax tables here. But, like the federal government, state governments provide tax tables and, also like the federal government, base the tax on an employee's gross pay, marital status, and withholding allowances.

For our example of Lang Kirkland, we will assume his state income tax for this pay period is $12.

FICA Taxes

The **Federal Insurance Contributions Act (FICA),** more commonly called **social security,** requires employers to withhold a certain percentage of each employee's earnings as a contribution to the federal social security program.

There are two parts to the FICA tax: (1) **OASDI** (old age, survivors, and disability insurance) and (2) **Medicare** (hospital insurance). Before 1991, employers could combine and report the FICA tax as a single amount. Starting in 1991, however, the two parts of the tax are calculated on different bases. For OASDI, the current rate is 6.2% of the first $68,400 earned in the year. For Medicare, the current rate is 1.45% of all earnings (no annual limit).

Now let's return to our example of Lang Kirkland. Lang's weekly earnings were $350 and he has not reached $68,400 in earnings for the year. So, his FICA tax for this payroll period is:

OASDI	$350 × .062	=	$21.70
Medicare	350 × .0145	=	5.08
Total			$26.78

Single Persons—Weekly Payroll Period

And the wages are—		And the number of withholding allowances claimed is —										
At least	But less than	0	1	2	3	4	5	6	7	8	9	10
		The amount of income tax to be withheld shall be—										
$ 0	$ 25	$ 0	$ 0	$ 0	$ 0	$ 0	$ 0	$ 0	$ 0	$ 0	$ 0	$ 0
25	30	1	0	0	0	0	0	0	0	0	0	0
30	35	2	0	0	0	0	0	0	0	0	0	0
35	40	3	0	0	0	0	0	0	0	0	0	0
40	45	3	0	0	0	0	0	0	0	0	0	0
45	50	4	0	0	0	0	0	0	0	0	0	0
50	55	5	0	0	0	0	0	0	0	0	0	0
55	60	6	0	0	0	0	0	0	0	0	0	0
60	65	6	1	0	0	0	0	0	0	0	0	0
65	70	7	1	0	0	0	0	0	0	0	0	0
70	75	8	2	0	0	0	0	0	0	0	0	0
75	80	9	3	0	0	0	0	0	0	0	0	0
80	85	9	4	0	0	0	0	0	0	0	0	0
85	90	10	4	0	0	0	0	0	0	0	0	0
90	95	11	5	0	0	0	0	0	0	0	0	0
95	100	12	6	0	0	0	0	0	0	0	0	0
100	105	12	7	1	0	0	0	0	0	0	0	0
105	110	13	7	2	0	0	0	0	0	0	0	0
110	115	14	8	3	0	0	0	0	0	0	0	0
115	120	15	9	3	0	0	0	0	0	0	0	0
120	125	15	10	4	0	0	0	0	0	0	0	0
125	130	16	10	5	0	0	0	0	0	0	0	0
130	135	17	11	6	0	0	0	0	0	0	0	0
135	140	18	12	6	1	0	0	0	0	0	0	0
140	145	18	13	7	1	0	0	0	0	0	0	0
145	150	19	13	8	2	0	0	0	0	0	0	0
150	155	20	14	9	3	0	0	0	0	0	0	0
155	160	21	15	9	4	0	0	0	0	0	0	0
160	165	21	16	10	4	0	0	0	0	0	0	0
165	170	22	16	11	5	0	0	0	0	0	0	0
170	175	23	17	12	6	0	0	0	0	0	0	0
175	180	24	18	12	7	1	0	0	0	0	0	0
180	185	24	19	13	7	2	0	0	0	0	0	0
185	190	25	19	14	8	3	0	0	0	0	0	0
190	195	26	20	15	9	3	0	0	0	0	0	0
195	200	27	21	15	10	4	0	0	0	0	0	0
200	210	28	22	16	11	5	0	0	0	0	0	0
210	220	29	24	18	12	7	1	0	0	0	0	0
220	230	31	25	19	14	8	3	0	0	0	0	0
230	240	32	27	21	15	10	4	0	0	0	0	0
240	250	34	28	22	17	11	6	0	0	0	0	0
250	260	35	30	24	18	13	7	1	0	0	0	0
260	270	37	31	25	20	14	9	3	0	0	0	0
270	280	38	33	27	21	16	10	4	0	0	0	0
280	290	40	34	28	23	17	12	6	0	0	0	0
290	300	41	36	30	24	19	13	7	2	0	0	0
300	310	43	37	31	26	20	15	9	3	0	0	0
310	320	44	39	33	27	22	16	10	5	0	0	0
320	330	46	40	34	29	23	18	12	6	1	0	0
330	340	47	42	36	30	25	19	13	8	2	0	0
340	350	49	43	37	32	26	21	15	9	4	0	0
350	360	50	45	39	33	28	22	16	11	5	0	0
360	370	52	46	40	35	29	24	18	12	7	1	0
370	380	55	48	42	36	31	25	19	14	8	3	0
380	390	58	49	43	38	32	27	21	15	10	4	0

And the wages are—		And the number of withholding allowances claimed is—										
At least	But less than	0	1	2	3	4	5	6	7	8	9	10
		The amount of income tax to be withheld shall be—										
180	185	19	13	7	2	0	0	0	0	0	0	0
185	190	19	14	8	2	0	0	0	0	0	0	0
190	195	20	14	9	3	0	0	0	0	0	0	0
195	200	21	15	10	4	0	0	0	0	0	0	0
200	210	22	16	11	5	0	0	0	0	0	0	0
210	220	23	18	12	7	1	0	0	0	0	0	0
220	230	25	19	14	8	2	0	0	0	0	0	0
230	240	26	21	15	10	4	0	0	0	0	0	0
240	250	28	22	17	11	5	0	0	0	0	0	0
250	260	29	24	18	13	7	1	0	0	0	0	0
260	270	31	25	20	14	8	3	0	0	0	0	0
270	280	32	27	21	16	10	4	0	0	0	0	0
280	290	34	28	23	17	11	6	0	0	0	0	0
290	300	35	30	24	19	13	7	2	0	0	0	0
300	310	37	31	26	20	14	9	3	0	0	0	0
310	320	38	33	27	22	16	10	5	0	0	0	0
320	330	40	34	29	23	17	12	6	1	0	0	0
330	340	41	36	30	25	19	13	8	2	0	0	0
340	350	43	37	32	26	20	15	9	4	0	0	0
350	360	44	39	33	28	22	16	11	5	0	0	0
360	370	46	40	35	29	23	18	12	7	1	0	0
370	380	47	42	36	31	25	19	14	8	2	0	0
380	390	49	43	38	32	26	21	15	10	4	0	0
390	400	50	45	39	34	28	22	17	11	5	0	0
400	410	52	46	41	35	29	24	18	13	7	1	0
410	420	53	48	42	37	31	25	20	14	8	3	0
420	430	55	49	44	38	32	27	21	16	10	4	0
430	440	56	51	45	40	34	28	23	17	11	6	0
440	450	58	52	47	41	35	30	24	19	13	7	2
450	460	59	54	48	43	37	31	26	20	14	9	3
460	470	61	55	50	44	38	33	27	22	16	10	5
470	480	62	57	51	46	40	34	29	23	17	12	6
480	490	64	58	53	47	41	36	30	25	19	13	8
490	500	65	60	54	49	43	37	32	26	20	15	9
500	510	67	61	56	50	44	39	33	28	22	16	11
510	520	68	63	57	52	46	40	35	29	23	18	12
520	530	70	64	59	53	47	42	36	31	25	19	14
530	540	71	66	60	55	49	43	38	32	26	21	15
540	550	73	67	62	56	50	45	39	34	28	22	17
550	560	74	69	63	58	52	46	41	35	29	24	18

DEDUCTIONS AGREED TO BY THE EMPLOYEE AND THE EMPLOYER

We have discussed the most common deductions that are required by law to be withheld. There are many other deductions that an employee can agree to have withheld from his or her pay. The more common deductions are health insurance, retirement plans, savings plans, loan repayments, union dues, and amounts to purchase U.S. government savings bonds. For our example of Lang Kirkland, we will assume he has $25 withheld weekly for savings bonds.

CALCULATING NET PAY

As stated earlier, net pay is the amount of an employee's check after all deductions have been made. Lang Kirkland's net pay is calculated as follows:

Gross pay	. . . .	$350.00
Less: Federal income tax	. . . .$28.00	
State income tax	. . . . .12.00	
FICA tax	. . . . .26.78	
Savings bonds	. . . . .25.00	
Total deductions	. . . .	91.78
Net pay	. . . .	$258.22

Self-Check

Marie Walters is single and claims one withholding allowance. Last week she earned $372. In addition to FICA and federal income taxes, she has $25 withheld for insurance. What is her net pay?

Solution

		$372.00
Income taxes	. . . .$48.00	
FICA—OASDI	. . . . .23.06	
FICA—Medicare	. . . . 5.39	
Insurance	. . . . .25.00	101.45
Net Pay	. . . .	$270.55

THE PAYROLL REGISTER

The **payroll register** is a form used by businesses to organize and summarize payroll data. The actual design of the payroll register depends on the needs of the individual company. The payroll register shown on page 380 is for Allied Products Company, a distributor of soft drinks.

FOR WEEK ENDING _June 29_ 20 _XX_

#	Employee's Name	MARITAL STATUS	NO. W.H. ALLOW.	REGULAR EARNINGS HOURS WORKED	RATE PER HOUR	AMOUNT	OVERTIME EARNINGS HOURS WORKED	RATE PER HOUR	AMOUNT	TOTAL EARNINGS	DEDUCTIONS FICA TAX	FEDERAL INCOME TAX	STATE INCOME TAX	OTHER	NET PAY
1	Adams, Phillip H.	M	3	40	8 00	320 00	—	—	—	320 00	24 48	23 00	14 20	4 00	254 32
2	Barker, Patrice R.	M	1	40	9 00	360 00	—	—	—	360 00	27 54	40 00	18 60	4 00	269 86
3	Darsky, Randolph P.	M	4	44	10 50	420 00	4	15 75	63 00	483 00	36 95	41 00	21 00	—	384 05
4	Franklin, Pamela R.	M	1	40	6 00	240 00	—	—	—	240 00	18 36	22 00	8 70	—	190 94
5	Helms, Faye	M	2	40	12 00	480 00	—	—	—	480 00	36 72	53 00	23 40	4 00	362 88
6	Purdue, Joy H.	M	2	46	6 00	240 00	6	9 00	54 00	294 00	22 49	24 00	13 00	15 00	220 51
7	Turner, Joseph J.	M	1	38	8 00	304 00	—	—	—	304 00	23 26	31 00	14 00	4 00	231 74
8	Veal, Michele	S	1	40	8 00	320 00	—	—	—	320 00	24 48	40 00	18 02	—	237 50
9	Watson, Teresa L.	M	1	40	6 00	240 00	—	—	—	240 00	18 36	22 00	8 70	4 00	186 94
10	Williams, Todd K.	S	1	40	9 00	360 00	—	—	—	360 00	27 54	46 00	22 90	—	263 56
11	Totals					3284 00				3401 00	260 18	342 00	161 52	35 00	2602 30
12															
13															
14															
15															
16															
17															
18															
19															
20															
21															

Name _____ Date_____

Exercises

1. Using the tax tables shown in this unit, calculate the amount of federal income tax to be withheld from the earnings of each of the following employees:

Employee	Marital Status	Withholding Allowances	Gross Pay	Amount of Income Tax
A. Dye	M	4	$380	$_____
S. Kaye	S	1	325	_____
B. Birch	M	2	402	_____
M. Man	S	1	290	_____
H. Book	M	5	425	_____
C. Hendrix	M	1	425	_____
M. Mallard	M	3	395	_____
T. O'Neal	S	1	389	_____

2. Calculate the federal income tax, state income tax, FICA tax, and net pay for each of the following employees. Assume a rate of 8% for state income tax. No employee has reached or exceeded $68,400 in earnings.

Employee	Martial Status	Withhold. Allowance	Gross Pay	Federal Income Tax	State Income Tax	Total FICA	Net Pay
L. Walk	M	4	$480	$_____	$_____	$_____	$_____
W. Mays	S	1	340	_____	_____	_____	_____
B. Little	M	5	390	_____	_____	_____	_____
D. Kramer	M	2	422	_____	_____	_____	_____
E. Hall	S	1	372	_____	_____	_____	_____

3. Calculate the FICA tax on each of the following employees:

Employee	Earnings Before This Pay Period	Earnings This Pay Period	FICA OASDI	Medicare
A	$12,800	$450	$_____	$_____
B	53,100	800	_____	_____
C	56,000	980	_____	_____
D	34,500	535	_____	_____
E	53,050	590	_____	_____

Applications

1. James Jordan is married and claims three withholding allowances. His weekly salary is $320. What is the amount of federal income tax to be withheld?

2. Sue Lesan is single and claims one withholding allowance. Last week her gross pay was $325. Compute her federal income tax, FICA tax, and net pay. Assume her state income tax is $12.

3. Glen Dyer's gross pay for the current pay period was $840. What is his OASDI tax if his earnings before this pay period were (a) $68,200 and (b) $41,900?

4. Danny Czonka is married and claims four withholding allowances. Last week his gross pay was $390 and his total earnings for the year are $21,500. Find his (a) federal income tax, (b) FICA tax, and (c) net pay, assuming he has a state income tax of $14 and a health insurance deduction of $24.

5. Linda Day's take-home pay for this week was $220. Her gross pay was $300. What percent of her gross pay was withheld? [*Hint:* P = $80 ($300 − $220), B = $300, Find R.]

ANSWERS

1. _____

2. _____

3. a. _____

   b._____

4. a. _____

   b._____

   c. _____

5. _____

6. Tom Posney's FICA tax for the week was $17.21, which was 7.65% of his gross pay. What was his gross pay? (*Hint:* P = $17.21, R = .0765, find B.)

7. The partial payroll register of Goozner Company is shown on page 385. Complete the payroll register.

ANSWERS

6. _____

7. _____

PAYROLL REGISTER

FOR WEEK ENDING __August 15__ 20 __X X__

	Employee's Name	MARITAL STATUS	NO. W.H. ALLOW.	REGULAR EARNINGS			OVERTIME EARNINGS			TOTAL EARNINGS	DEDUCTIONS				NET PAY
				HOURS WORKED	RATE PER HOUR	AMOUNT	HOURS WORKED	RATE PER HOUR	AMOUNT		FICA TAX	FEDERAL INCOME TAX	STATE INCOME TAX	OTHER	
1	Abrahms, Polly	M	2	40	9 00		2						8 00	6 00	
2	Adams, Ray	M	4	40	10 00		8						10 40		
3	Conners, Timothy	M	1	40	8 00		1						6 40		
4	Denton, Robert	S	0	40	7 00		4						6 44		
5	Grey, Taylor	M	2	40	11 00		1						8 80	6 00	
6	Jones, Kaye	S	1	40	9 50		1						7 60		
7	Kitchens, Wylene	M	1	40	11 00		1						8 90		
8	Mays, Wanda	M	4	40	12 00		1						9 60		
9	Patterson, Lane	S	1	40	9 00		1						7 20		
10	Roberts, Fred	M	2	40	10 00		8						10 08	6 00	
11	Totals												83 42	18 00	
12															
13															
14															
15															
16															
17															
18															
19															
20															
21															

Work Space

SECTION VI Math for Consumer and Business Use

unit 6.5

Finding Commissions

After completing Unit 6.5, you will be able to:

1. Compute commissions in different situations.

Instead of getting an hourly wage or a fixed salary, many salespeople work on a commission basis; that is, they get a percentage of the value of goods and services they sell. For example, Lisa Todd is a real estate salesperson. Lisa does not get a salary; instead, she is paid 5% of the price of each house she sells. Last week Lisa sold a house for $60,000. Her 5% commission is calculated as follows:

$$
\begin{array}{ll}
\$60,000 & \text{price of house} \\
\times \quad\underline{.05} & \text{commission rate} \\
\$\ 3,000 & \text{amount of commission}
\end{array}
$$

In commission problems, you are actually working with the formula $B \times R = P$. The total selling price of the goods or services is B, the rate of commission is R, and P is the result of multiplying $B \times R$. We can illustrate this with a pie diagram:

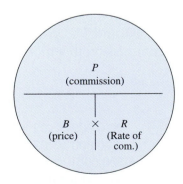

Let's look at another example.

Example

John Distasio works for a large travel agency. John receives a 6% commission on all overseas vacations he sells. Last week John sold a European vacation to a couple for $3,000. What was his commission?

Solution

$$\$3,000 \times .06 = \$180$$

Some salespeople are paid on a salary plus commission basis. For example, Lynn Haden earns a salary of $200 a week plus a 3% commission on all merchandise she sells. Last week her total sales amounted to $4,000. Her weekly earnings are:

$$
\begin{array}{ll}
\$200 & \text{salary} \\
+\ \underline{120} & \text{commission (\$4,000} \times .03) \\
\$320 & \text{total earnings}
\end{array}
$$

Commissions are sometimes figured on a sliding scale. Let's look at a couple of examples to see how a sliding scale works.

Example 1

Wendy Carpenter, a salesperson, gets a 4% commission on all sales up to $50,000 and a 6% commission on all sales over $50,000. Last month Wendy's total sales amounted to $62,000. What was her commission?

Solution

Step 1: Figure commission on the first $50,000:

$$\$50,000 \times .04 = \$2,000$$

Step 2: Figure commission on sales over $50,000:

$$\$62,000 - \$50,000 = \$12,000 \times .06 = \$720$$

Step 3: Add the two amounts:

$$\$2,000 + \$720 = \mathbf{\$2,720}$$

Example 2

Donald Haick gets 2% commission on the first $10,000 of goods sold each month, 3% on the next $10,000, and 5% on any goods sold in excess of $20,000. What is his commission in a month in which his sales totaled $34,000?

Solution

$$
\begin{array}{l}
\$10,000 \times .02 = \$200 \\
\$10,000 \times .03 = \$300 \\
\$14,000\ (\$34,000 - \$20,000) \times .05 = \$700 \\
\$200 + \$300 + \$700 = \mathbf{\$1,200\ total\ commission}
\end{array}
$$

Work Space

unit 6.5

Exercises

1. The following salespeople are paid a commission based on their monthly sales. Find the commission of each salesperson:

Salesperson	Rate of Commission	Monthly Sales	Commission
John Lang	8%	$44,000	$_____
Jane Lewis	6%	30,800	_____
Todd King	$4\frac{1}{2}$%	45,000	_____
Lee Cornwell	7%	18,000	_____
Jesse Fazio	$8\frac{1}{2}$%	24,000	_____
Li Smitherman	$3\frac{3}{4}$%	12,000	_____
Karl Schwartz	10%	35,600	_____
Lou Kingrey	5%	43,546	_____

2. The following salespeople are paid a salary plus a commission on their weekly sales. Find the total earnings of each salesperson:

Salesperson	Salary	Rate of Commission	Weekly Sales	Total Earnings
Dave Rodgers	$225	2%	$8,000	$_____
Sueanne Lee	340	3%	6,200	_____
Randy Jones	300	5%	3,400	_____
Earl Hillary	275	$4\frac{1}{2}$%	2,500	_____
Lori Pope	350	6%	5,000	_____
Ann Foreman	230	7%	6,944	_____
Hain Bach	500	$1\frac{1}{2}$%	6,000	_____
Jay Whitmore	190	10%	1,250	_____

3. The following salespeople are paid a salary plus a stated commission of all weekly sales over $4,000. Find the total earnings of each salesperson:

Salesperson	Salary	Rate of Commission	Total Weekly Sales	Earnings
Taylor Dawn	$300	6%	$8,200	$_____
Alice Ingrum	225	4%	6,100	_____
Henry Jordan	300	5%	6,500	_____

Salesperson	Salary	Rate of Commission	Total Weekly Sales	Earnings
Paul Corelli	295	6%	6,800	_____
Louis Crowell	235	4%	5,600	_____
Wendy Cars	325	5%	4,900	_____
Lisa Dobbs	345	7%	5,675	_____

Name _____ **Date**_____

Applications

1. Tim Reed is a salesperson for Harvey Lumber Company. Tim gets a salary of $325 weekly plus a 4% commission on his total weekly sales. This week his total sales amounted to $9,200. What are his total earnings?

1. _____

2. _____

3. _____

4. _____

2. Renee Friedman receives a monthly salary of $1,200 plus a 6% commission on monthly sales in excess of $12,000. Last month her sales totaled $14,700. What are her total earnings?

3. Randy Knight works on a straight commission basis. His commission is paid on a sliding scale, as follows:

   2% on first $10,000 sold
   3% on next $15,000 sold
   5% on next $15,000 sold
   8% on all sales in excess of $40,000

   Last month Randy's total sales amounted to $47,500. What is his commission?

4. Joanne Preston's 6% commission for selling a house came to $3,600. What was the selling price of the house? (*Hint:* $R = 6\%$, $P = \$3,600$, find B.)

5. Pylant Products Company pays its salespeople a base salary and a commission based on the following sliding scale: 5% of the first $1,000, 8% of the next $1,000, and 10% of all sales above $2,000. Complete the following earnings report:

PYLANT PRODUCTS COMPANY
Monthly Earnings Report
For Month Ended July 31, 20–

Salesperson	Total Sales	Amount of Commission	Base Salary	Total Earnings
Lynn Porter	$4,800	$_____	$1,000	$_____
Jan Herdanez	2,450	_____	1,200	_____
Dave Stover	3,500	_____	1,490	_____
Kay Segram	2,300	_____	1,450	_____
Lyle Kirkland	4,560	_____	1,250	_____
Anne Moore	6,700	_____	1,425	_____
Vickie Cooke	4,200	_____	1,230	_____
Totals		$_____		$_____

Business Inventories

After completing Unit 6.6, you will be able to:

1. Differentiate between a periodic inventory system and a perpetual inventory system.
2. Determine the dollar value of inventory.

An **inventory** is the quantity of goods remaining on hand at the end of a period of time. For example, the inventory of Sears would be the total amount of goods (merchandise) on hand and unsold at the end of the year.

There are two systems used to determine the quantity of goods remaining on hand: the **periodic inventory system** and the **perpetual inventory system.** Under the periodic system, goods are physically counted at the end of a period of time, such as a month or a year. When all goods have been counted, a dollar value will be assigned to the inventory (we will discuss how a value is assigned in the next section of this unit).

Under the perpetual system, written records are maintained that show a continuous or "running" inventory balance for each item sold by the business. When new items arrive, they are added to the inventory records. When items are sold, the number sold is subtracted from the inventory records. In that way, the inventory records are always up-to-date and constantly show the amount of goods on hand. An example of a perpetual inventory record looks like this:

Perpetual Inventory Record

Item	Silk Ties		Stock Number	7007		Reorder Point	75

				Balance on Hand		
Date	Purchase	Unit Cost	Sold	Unit Cost	Quantity	Total
May 1				$18.00	75	$1,350
2	400	$18.00		18.00	475	8,550
5			180	18.00	295	5,310
8			125	18.00	170	3,060
12			40	18.00	130	2,340
21			12	18.00	118	2,124
29			15	18.00	103	1,854
31	400	18.50		18.00	103	
				18.50	400	9,254

 ## ASSIGNING A DOLLAR VALUE TO INVENTORY

Under the perpetual system, as we just discussed, you keep a running value of each inventory item. Under the periodic system, you wait until the end of the period, count the goods on hand, and then assign a value to the goods on hand. There are several acceptable methods of assigning a value to inventory. The most common methods are specific identification; weighted average; first-in, first-out (FIFO); and last-in, first-out (LIFO). Each method is discussed next.

Specific Identification

Under the **specific identification** method, each item on hand is "specifically" identified with its actual cost. Income items are often tagged to indicate their cost and the date on which they were bought. Then, at the end of the period, each item is counted and its unit cost is recorded. The costs of all items are then added to get the total cost of the inventory.

Example

Gene Sterling is a used auto dealer. At the end of June, Gene's records show the following:

Vehicle Number	Cost
76	$ 7,400
105	6,280
119	5,495
204	7,210
207	3,860
219	11,240
255	9,130

Calculate the value of Gene's inventory.

Solution

$7,400 + $6,280 + $5,495 + $7,210 + 3,860 + $11,240 + $9,130 = **$50,615**

It was easy for Gene to determine the value of his inventory because he only had seven cars on hand at the end of the month. But, consider the following data for Barbara O'Malley, owner of a small jeans store:

Date	Quantity Bought	Unit Cost
July 3	20	$18.00
16	40	18.50
24	10	18.75
31	50	18.90

At the end of July, Barbara counts her unsold jeans and finds that 25 pairs are on hand. By the codes she entered on the labels at the time of purchase, she can tell that 18 of the 25 are from the July 31 purchase and the remaining are from July 16. Thus, she determines the value of her inventory as follows:

$$18 \times \$18.90 = \$340.20$$
$$7 \times \ 18.50 = \underline{\ 129.50}$$
$$\text{Total} \qquad \$469.70$$

The specific identification method works well when a business has a small inventory of easily identified items. However, it is usually too time-consuming for businesses with large inventories.

The Weighted Average Method

The **weighted average method** works well for large inventories. Under this method, an average price per unit is computed for all goods in inventory. The value of the inventory is then computed by multiplying the number of units on hand at the end of the period by the average cost per unit. Let's look at an example of how it works.

Example

On September 30, The Hobby Store showed the following data for one of its video games, *Chi's Quest*.

Date	Description	Quantity	Unit Cost
Sept. 1	Beginning inventory	30	$10.00
8	Purchase	40	10.50
15	Purchase	40	10.75
26	Purchase	25	11.00

On September 30, a count of inventory shows that 29 units are on hand. Compute the value of the inventory.

Solution

Step 1: Determine the cost of goods available for sale:

		Units	×	Unit Cost	=	Total Cost
Sept. 1	Beginning inventory	30	×	$10.00	=	$ 300
8	Purchase	40	×	10.50	=	420
15	Purchase	40	×	10.75	=	430
26	Purchase	25	×	11.00	=	275
Units available for sale		135				$1,425

Step 2: Determine the average cost per unit:

$$\frac{\$1,425}{135} = \$10.56 \quad \text{average cost per unit}$$

Step 3: Apply the average cost per unit to the number of units remaining on hand:

$$\$10.56 \times 29 \text{ units} = \$306.24 \qquad \text{inventory value}$$

First-in, first-out (FIFO)

First-in, first-out (FIFO) is a popular method of assigning a value to inventory when specific identification is not possible. FIFO is *an assumed cost flow* method. This means that goods are assumed to flow (or move) in a certain direction, and a value is assigned to goods based on that assumed flow. For example, if your inventory included perishable goods such as milk, eggs, fruit, vegetables, and so on, you would naturally attempt to sell the older items first. FIFO assumes this flow. That is, the first goods acquired are the first goods assumed to be sold. As a result, the ending inventory is assumed to be made up of the last goods acquired. To illustrate this, let's return to our example of Chi's Quest video games. Using FIFO, we can assign a value to the 29 games left on hand at the end of September as follows:

Description	No. of Units		Unit Cost		Total Cost
From Sept. 26 purchase	25	×	$11.00	=	$275
From Sept. 15 purchase	4	×	10.75	=	43
Ending inventory	29				$318

The 25 units purchased on September 26 are assumed to be on hand, and 4 of the units purchased on September 15 are assumed to be on hand. All other units are assumed to be sold. Since most businesses attempt to sell old goods before new goods, FIFO is generally in harmony with the actual movement of goods in a business.

Last-in, First-out (LIFO)

The **last-in, first out (LIFO)** method is also an assumed cost flow method. LIFO assumes the last goods acquired are the first goods to be sold. Thus, the ending inventory is assumed to be part of the earliest goods acquired. To illustrate LIFO, let's return once again to the 29-item count of Chi's Quest video games. Since

LIFO assumes inventory is made up of the earliest costs, the 29 units on hand on September 30 are assigned a valued based on the value of the September 1 beginning inventory:

$$29 \text{ units} \times \$10.00 = \$290$$

In most businesses, the LIFO method does not parallel the actual movement of goods. Keep in mind, however, that this is a method of assigning a value to inventory—not a method of counting the units on hand. As a result, any business can use the LIFO method. Each individual business must look at its particular needs and decide on which inventory method works best for it.

Self-Check

Sterling Music Store shows the following for this year:

Beginning inventory, January 1 . . . .   300 units @ $7.00
Purchase, March 1 . . . .   700 units @ 7.20
Purchase, June 1 . . . . 1,200 units @ 7.25
Purchase, September 1 . . . . 1,600 units @ 7.30
Purchase, October 1 . . . .   200 units @ 7.40

On December 31, a physical count revealed that 400 units remained on hand. Compute the value of the 400 units under each of the following methods:

(a) Specific identification, assuming 100 of the units were from the January 1 inventory, 100 were from the June 1 purchase, and 200 were from the October 1 purchase.
(b) Weighted average
(c) FIFO
(d) LIFO

Solution

(a) Specific identification

$$
\begin{aligned}
100 \times \$7.00 &= \$\ \ 700 \\
100 \times \ \ 7.25 &= \ \ \ \ 725 \\
200 \times \ \ 7.40 &= \underline{\ 1,480} \\
&\ \ \ \$2,905
\end{aligned}
$$

(b) Weighted average

No. of Units	×	Unit Cost	=	Total Cost
300	×	$7.00	=	$ 2,100
700	×	7.20	=	5,040
1,200	×	7.25	=	8,700
1,600	×	7.30	=	11,680
200	×	7.40	=	1,480
4,000 units		Totals		$29,000

$29,000 ÷ 4,000 units = $7.25 average cost per unit
400 units × $7.25 = $2,900

(c) FIFO
Remember:

FIFO starts with the latest costs.	→	200 units × $7.40 = $1,480
		200 units × 7.30 = 1,460
		$2,940

(d) LIFO

LIFO starts with the earliest costs.	→	300 units × $7.00 = $2,100
		100 units × 7.20 = 720
		$2,820

NOTES

1. The beginning inventory balance is always included in the weighted average method.
2. The beginning inventory—not the first purchase—is the first item considered in the LIFO method. This always holds true when there is a beginning inventory.

Name _____ Date _____

Exercises

Keisha Swain's inventory of top coats shows the following:

Balance, January 1 .... 60 units @ $60
Purchase, March 1 .... 80 units @ $61
Purchase, May 1 .... 90 units @ $62
Purchase, July 1 .... 40 units @ $63
Purchase, October 1 .... 30 units @ $64
Balance on hand, December 31 .... 45 units

Compute the value of her ending inventory using each of the following methods:

1. Specific identification, assuming 30 of the units are from January, 10 are from March, and 5 are from October.

2. Weighted average

3. FIFO

4. LIFO

From the following inventory data, compute the number of units remaining in inventory and the cost of the inventory using each of the following methods: weighted average, FIFO, and LIFO:

Balance, January 1 . . . .   8,000 units @ $4.25
Purchase, May 2 . . . . 12,000 units @ $4.30
Purchase, July 6 . . . .   9,000 units @ $4.55
Purchase, October 4 . . . . 11,500 units @ $4.60
Purchase, December 12 . . . .   7,800 units @ $4.70

48,300

Number of units sold 38,200

5. Number of units remaining in inventory = _____

6. Weighted average inventory cost = _____

7. FIFO inventory cost = _____

8. LIFO inventory cost = _____

unit 6.6

Applications

1. Sam King's inventory of electric guitars consists of six items valued as follows: #472, $625.00; #495, $837.50; #510, $415.00; #711, $360.00; #712, $522.50; and #803, $870.00. Calculate the value of her inventory of guitars:

ANSWERS

1. _____

2. a. _____

   b. _____

   c. _____

2. Compute the weighted average unit price in each of the following cases:

  a. 40 units @ $6
    20 units @ $9

  b. 100 units @ $7.10
    200 units @ $7.20
    200 units @ $7.30

  c. 50 units @ $9.00
    170 units @ $9.30
    210 units @ $9.60
    70 units @ $9.90

3. Lori Hume began the month with 30 auto batteries valued at $40 each. She then purchased 20 more batteries during the month at a cost of $45 each. At the end of the month, 10 units remained in her stockroom. Compute the ending inventory value by the weighted average method.

4. Alli Heart purchased 37 coats on July 1 at $29.95 each and 50 more on July 22 at $28.40 each. If 15 coats remain on hand at the end of the month of July, what is their value using the FIFO method?

5. Todd Lane began the month of March with a supply of 19 notebooks at $.22 each. During March, he made two purchases: 106 at $.23 and 32 at $.24. If 45 remain at the end of March, determine their value using the FIFO method.

Computing Depreciation by Straight-Line and Units of Production Methods

After completing Unit 6.7, you will be able to:

1. Compute depreciation by the straight-line method.
2. Compute depreciation by the units of production method.

Assets are items of value owned by a business. Examples include cash, supplies, cars, equipment, inventory, furniture, buildings, and land.

Some assets, such as supplies, only last for a short period of time—usually less than a year. Other assets can remain useful for many years. These assets are called **long-term assets** and include cars, buildings, equipment, and land.

With the exception of land, all long-term assets eventually wear out or become obsolete. As an asset wears out or becomes obsolete, it loses its usefulness. This loss of usefulness is called **depreciation.** Most of us have experienced depreciation through the ownership of personal assets. If you have ever owned a car, for example, you know it is worth less every year you own it.

In this unit, we will discuss two common ways of computing depreciation: the straight-line method and the units of production method. In Unit 6.8, we will discuss two other common depreciation methods: the declining-balance method and the sum-of-the-years' digits method.

STRAIGHT-LINE METHOD

The most popular method for computing depreciation has traditionally been the **straight-line method.** To compute depreciation by the straight-line method, you need three things:

1. The cost of an asset
2. The estimated useful life of the asset
3. The salvage value, or scrap value, of the asset—the amount you think the asset will be worth at the end of its useful life

The straight-line method is based on the following formula:

$$\frac{\text{Cost} - \text{salvage value}}{\text{Estimated useful life}} = \text{annual depreciation}$$

For example, if an asset has a cost of $12,000, an estimated salvage value of $2,000, and an estimated useful life (EUL) of 5 years, the straight-line depreciation is:

$$\frac{C - S}{\text{EUL}} = \frac{\$12,000 - \$2,000}{5 \text{ years}} = \frac{\$10,000}{5} = \$2,000 \text{ a year}$$

The straight-line method gives an equal amount of depreciation for each full year the asset is used. For this example, the asset is estimated to depreciate $2,000 each year the asset is used. Let's look at another example.

Example

Find the annual depreciation expense for an asset with a cost of $16,000, an estimated salvage value of $1,000, and an EUL of 10 years.

Solution

$$\frac{C - S}{\text{EUL}} = \frac{\$16,000 - \$1,000}{10 \text{ years}} = \frac{\$15,000}{10} = \$1,500 \text{ a year}$$

Self-Check

Find the annual depreciation expense for an asset with a cost of $12,000, an estimated salvage value of $3,000, and an EUL of 3 years.

Solution

$$\frac{C - S}{\text{EUL}} = \frac{\$12,000 - \$3,000}{3 \text{ years}} = \frac{\$9,000}{3} = \$3,000 \text{ a year}$$

≋ THE UNITS OF PRODUCTION METHOD

The **units of production method** uses the same formula as the straight-line method, except that the life of the asset is not stated in years; it is stated in terms of productive output. By productive output we mean what the asset is expected to produce. For example, you can estimate the life of a truck in terms of how many miles it will be driven. Likewise, you can estimate the life of a tractor in terms of how many hours it will operate. To illustrate the units of production method, assume the following about a truck:

Cost $25,000
Estimated salvage value . . . . $ 5,000
Estimated useful life . . . . 100,000 miles

We compute a depreciation rate per mile as follows:

$$\frac{C - S}{EUL} = \frac{\$25,000 - \$5,000}{100,000 \text{ miles}} = \frac{\$20,000}{100,000} = \$.20 \text{ per mile}$$

Now, if the truck is driven 30,000 miles this year, the depreciation would be:

$$
\begin{array}{r}
30{,}000 \\
\times \ \$ \quad .20 \\
\hline
\$6{,}000
\end{array}
$$

If the truck is driven 35,000 miles the next year, the depreciation would be $7,000 (35,000 miles × $.20).

Self-Check

A machine with a cost of $9,000 and an estimated salvage value of $1,000 is expected to produce 20,000 units. Find (a) the depreciation rate per unit of output, (b) the depreciation the first year of use assuming 4,000 units were produced, and (c) the depreciation the second year assuming 5,000 units were produced.

Solution

(a) $\dfrac{C - S}{EUL} = \dfrac{\$9{,}000 - \$1{,}000}{20{,}000 \text{ units}} = \dfrac{\$8{,}000}{20{,}000} = \$.40 \text{ per unit}$

(b) $4{,}000 \times \$.40 = \$1{,}600$

(c) $5{,}000 \times \$.40 = \$2{,}000$

unit
6.7

Exercises

1. What is an asset?

2. On an asset with an annual depreciation of $5,000, how much depreciation has accumulated after $4\frac{1}{2}$ years?

3. What factors are needed to calculate depreciation?

4. What is meant by "productive life"?

5. How does the formula for the units of production method differ from the straight-line method?

6. Compute straight-line depreciation for each of the following assets:

Asset	Cost	Estimated Salvage Value	Estimated Useful Life	Depreciation
a.	$20,000	$6,000	4 years	$_____
b.	6,000	–0–	10 years	_____
c.	12,000	2,000	5 years	_____

7. Compute the depreciation rate per unit for each of the following assets:

Asset	Cost	Estimated Salvage Value	Estimated Useful Life	Depreciation per Unit
a.	$15,000	$3,000	90,000 hours of operation	$_____
b.	40,000	8,000	50,000 miles driven	_____
c.	90,000	–0–	100,000 units produced	_____

unit 6.7

Applications

1. The Macon Company purchased a drill press on January 4, 20X1. The cost of the asset was $18,000, and the company estimates that it will last for 5 years and have a salvage value of $3,000. Using the straight-line method, compute the depreciation for the first 3 years of the asset's life.

2. Tillman Company uses the units of production method. A machine costing $36,000 with an estimated life of 105,000 units and salvage value of $4,500 has the following production figures:

First year . . . . .28,300 units
Second year . . . . . . .30,000 units
Third year . . . . .24,500 units
Fourth year . . . . .26,000 units

Compute the depreciation for each year.

3. An asset with a cost of $14,500 and an estimated salvage value of $1,300 is to be depreciated by the straight-line method over 11 years. Find (a) the depreciation for the first year and (b) the total amount of depreciation after 5 years.

3. _____

4. _____

4. Using the units of production method, compute the depreciation per year for the following truck:

Cost $52,000
Estimated salvage value $10,000
Estimated useful life .... .200,000 miles
Miles driven:
    first year .... .28,000
    second year .... .65,000
    third year .... .52,000
    fourth year .... .43,000
    fifth year .... .12,000

unit 6.8

Computing Depreciation by Declining-Balance and Sum-of-the-Years' Digits Methods

After completing Unit 6.8, you will be able to:

1. Compute depreciation by the declining-balance method.
2. Compute depreciation by the sum-of-the-years' digits method.

In Unit 6.7, we learned how to compute depreciation using the straight-line method and the units of production method. In this unit, we are going to learn two other methods: the **declining-balance method** and the **sum-of-the years' digits method.** These methods are referred to as **accelerated depreciation methods** because they yield a larger amount of depreciation in the early years of an asset's life and less as the asset ages.

[handwritten note: Logic or assumption behind this : certain assets (eg, cars) depreciate more quickly in early years than in later years.?]

≈ DECLINING-BALANCE METHOD

The declining-balance method applies a constant rate to the declining value of an asset. The rate is found as follows:

Step 1. Divide 100% by the estimated useful life of the asset.

Step 2. Double the rate obtained in Step 1.

When you divide 100% by the life of an asset (Step 1), you get the *straight-line rate*. To get the declining-balance rate, the straight-line rate is doubled for most assets. In fact, this method is often called the *double declining-balance method*. To illustrate, assume an asset purchased for $10,000 has an estimated salvage value of $2,000, and an estimated useful life (EUL) of 5 years. First we compute the declining-balance rate, as follows:

Step 1:

$$\frac{100\%}{5 \text{ years}} = .20$$

Step 2:

$$.20 \times 2 = .40 \ (40\%)$$

Now we compute depreciation for the first year by multiplying the cost of the asset by the declining-balance rate, as follows:

You do not consider the salvage when using the declining-balance method.

$$\$10,000 \times .40 = \$4,000$$

To compute depreciation for the second year, we subtract the first year's depreciation ($4,000) from the cost of the asset ($10,000) to obtain its **book value** ($10,000 − $4,000 = $6,000 book value). We then multiply the book value by .40, as follows:

$$
\begin{array}{r}
\$10,000 \\
-\ \ \underline{4,000} \\
\$\ 6,000
\end{array}
\ \times .40 = \$2,400 \text{ depreciation for the second year}
$$

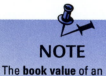

NOTE

The **book value** of an asset is its cost less the accumulated amount of depreciation.

Now, to compute depreciation for the third year, compute a new book value and multiply it by .40:

$$
\begin{array}{r}
\$6,000 \\
-\ \ \underline{2,400} \\
\$3,600
\end{array}
\ \times .40 = \$1,440 \text{ depreciation for the third year}
$$

Depreciation for the fourth year is found as follows:

$$
\begin{array}{r}
\$3,600 \\
-\ \ \underline{1,440} \\
\$2,160
\end{array}
\ \times .40 = \$864 \text{ depreciation for the fourth year}
$$

For the fifth year, we need to be a little careful because even though you do not consider the asset's salvage value in the calculation, <mark>you cannot depreciate the asset below its salvage value.</mark> To illustrate, let's look at the fifth year's calculation:

$$
\begin{array}{r}
\$2,160 \\
-\ \ \underline{864} \\
\$1,296
\end{array}
\ \times .40 = \$519
$$

Now, by computing the final book value, we will see that it has dropped below the $1,000 salvage value:

$$
\begin{array}{r}
\$1,296 \\
-\ \ \underline{519} \\
\$\ 777
\end{array}
\text{ final book value}
$$

Book value can't go below salvage value

Since the final book value would be below $1,000, we compute the final year's depreciation as follows:

$$
\begin{array}{rl}
\$1,296 & \text{book value at the end of the fourth year} \\
-\ \underline{1,000} & \text{asset's estimated salvage value} \\
\$\ 296 & \text{depreciation needed for the last year}
\end{array}
$$

Now, the final book value equals the estimated salvage value:

$1,296   book value at the end of the fourth year
  296   depreciation for the fifth year
$1,000   final book value

Self-Check

An asset has a cost of $12,000, an estimated salvage value of $1,000, and an estimated useful life of 4 years. Compute depreciation for each of the years using the declining-balance method.

Solution

$$\frac{100\%}{4 \text{ years}} = .25 \times 2 = .50 \ (50\%)$$

$12,000
×   .50
$ 6,000 1st year

$12,000
− 6,000
$ 6,000
×   .50
$ 3,000 2nd year

$6,000
− 3,000
$3,000
×   .50
$1,500 3rd year

$3,000
− 1,500
$1,500

$1,500
− 1,000
$ 500 4th year

SUM-OF-THE-YEARS' DIGITS METHOD

With the sum-of-the years' digits method, a series of fractions are multiplied by the cost of an asset less its estimated salvage value. The numerators of the fraction are the individual years in the asset's EUL. For example, if an asset is expected to last for 5 years, there will be a *first year*, a *second year*, a *third year*, a *fourth year*, and a *fifth year* in the life of the asset. We arrange these digits in reverse order to get our numerators: 5, 4, 3, 2, 1.

Then we sum the digits to get our denominator: 5 + 4 + 3 + 2 + 1 = 15. Thus, the first year's fraction will be $\frac{5}{15}$, the second year's fraction will be $\frac{4}{15}$, the third year's fraction will be $\frac{3}{15}$, and so on. Let's look at an example.

Example

An asset has a cost of $10,000, an estimated salvage value of $2,000, and an EUL of 4 years. Find the depreciation for each year by the sum-of-the years' digits method.

Solution

Sum of digits = 4 + 3 + 2 + 1 = 10

You consider salvage value with this method: $10,000 − $2,000 = $8,000.

First year: $\frac{4}{10}$ × $8,000 = $3,200

Second year: $\frac{3}{10}$ × $8,000 = 2,400

Third year: $\frac{2}{10}$ × $8,000 = 1,600

Fourth year: $\frac{1}{10}$ × $8,000 = 800

Self-Check

An asset has a cost of $15,000, an estimated salvage value of $3,000, and an EUL of 3 years. Find the depreciation for each year by the sum-of-the-years' digits method.

Solution

Sum of digits: 3 + 2 + 1 = 6

First year: $\frac{3}{6}$ × $12,000 = $6,000

Second year: $\frac{2}{6}$ × $12,000 = 4,000

Third year: $\frac{1}{6}$ × $12,000 = 2,000

Name _____ Date_____

Exercises

1. For each of the following lives of assets, find the declining-balance rate:

    a. 10 years _____   **b.** 6 years _____   **c.** 8 years _____

    d. 15 years _____   **e.** 20 years _____   **f.** 22 years _____

2. Find the sum-of-the-years' digits for each of the following assets:

    a. 6 years _____   **b.** 10 years _____   **c.** 8 years _____

3. Complete the following:

	Asset Cost	Salvage Value	Useful Life	Depreciation Method	First Year's Depreciation	Second Year's Depreciation	Third Year's Depreciation
a.	$12,000	$2,000	4 years	Declining-balance	$_____	$_____	$_____
b.	17,500	2,500	5 years	Sum-of-the years' digits	_____	_____	_____
c.	14,000	2,000	10 years	Declining-balance	_____	_____	_____

Applications

Name _____ Date_____

1. An asset with a cost of $18,000 and an estimated salvage value of $3,000 is to be depreciated over a 4-year period. Find the depreciation for each year using the declining-balance method.

1. _____

2. _____

3. _____

2. A transfer truck has a cost of $65,000, a salvage value of $15,000, and an estimated useful life (EUL) of 10 years. Compute the depreciation for each of the first 3 years using:

	Year 1	Year 2	Year 3
a. The declining-balance method:	$_____	$_____	$_____
b. The sum-of-the-years' digits method;	_____	_____	_____

3. Under the declining-balance method in problem 2, compute the book value of the truck after (a) the first year and (b) the third year.

4. A delivery van has a cost of $15,000, an estimated salvage value of $3,000, and an EUL of 5 years. Complete the following schedule assuming the use of the declining-balance method:

ANSWERS

4. _____

Year	Annual Depreciation	Accumulated Depreciation	Book Value
1	$_____	$_____	$_____
2	_____	_____	_____
3	_____	_____	_____
4	_____	_____	_____
5	_____	_____	_____

Sales Taxes

After completing Unit 6.9, you will be able to:

1. Determine the amount of sales tax on a retail sale.

Like private businesses, the various forms of governments (federal, state, and local) must have money (revenue) in order to operate. One source of operating revenue for state governments is the **sales tax,** which is a tax on the retail price of goods (and some services).

It is the seller's responsibility to collect the sales tax from the buyer and to pay it to the government. Sales tax rates vary from state to state and from county to county because it is also common for local governments to levy (charge) a sales tax. Goods subject to the tax also vary from state to state. For example, in some states, there is no sales tax on food items. For our problems, we will assume a sales tax rate of 7% and all items sold at retail are subject to the tax.

To calculate sales tax on whole dollar amounts, multiply the sales tax rate times the amount of sale.

Example
Find the sales tax on a $500 item in an area with a 7% sales tax rate.

Solution

$$
\begin{array}{rl}
\$500 & \text{amount of sale} \\
\times \ \underline{\ \ .07} & \text{sales tax rate} \\
\$\ 35 & \textbf{sales tax}
\end{array}
$$

To calculate sales tax on sales less than $1.00, use the following table:

Tax Rate is "seven cents on the dollar." So it stands to Reason that amts under a dollar wd be taxed at different rates. But why use a table? Multiplying Fractional amts by the tax Rate & Rounding up or down produces the same Result....

7% Sales Tax Table	
Amount of Sale	**Tax**
$0.00–.09	$.00
.10–.20	.01
.21–.33	.02
.34–.47	.03
.48–.62	.04
.63–.76	.05
.77–.91	.06
.92–.99	.07

Example 1

What is the sales tax on a sale of $.69?

Solution   *.69(.07) = .0483 = .05*

The amount of the sale, $.69, falls in the bracket $.63–$.76. Thus, the tax is $.05.

Example 2   *.95(.07) = .0665 = .07*

What is the sales tax and the total amount due on a sale of $.95?

Tax = $.07; $.95 + .07 = $1.02 amount due

To calculate the sales tax on sales that involve dollar and cent amounts, figure 7% on the whole dollar amounts and use the table for sales under $1.00.

Example 1

Find the sales tax on a sale of $10.50.

Solution

Step 1: Find the tax on the whole dollar amount:

$$\$10 \times .07 = \$.70$$

Step 2: Use the table to find the tax on $.50:   *10.50(.07) = .735 = .74*

$$\$.48 - .62 = \$.04$$

Step 3: Add the two amounts:

$$\$.70 + \$.04 = \mathbf{\$.74}$$

Example 2
Find the sales tax on a sale of $125.81.

Solution

$$\begin{aligned}
\$125 \times .07 &= \$8.75 \\
\text{Tax on \$.81 (table)} &= \underline{.06} \\
\text{Total tax} & \quad \mathbf{\$8.81}
\end{aligned}$$

Self-Check

Find the sales tax on a sale of $12.50 if the area has a 7% rate.

Solution

$.88

Name _____ Date_____

Exercises

Find the sales tax and the total amount due for each of the following sales. All sales are subject to a 7% sales tax. Use the table presented in the unit where needed.

	Amount of Sale	Amount of Sales Tax	Total Amount Due
1.	$ .69	$_____	$_____
2.	.99	_____	_____
3.	.25	_____	_____
4.	1.29	_____	_____
5.	12.95	_____	_____
6.	139.95	_____	_____
7.	525.00	_____	_____

Amount of Sale	Amount of Sales Tax	Total Amount Due
8. 900.00	_____	_____
9. 234.58	_____	_____
10. 107.99	_____	_____
11. 4,000.00	_____	_____
12. 3,569.69	_____	_____
13. 6,255.84	_____	_____
14. 8,002.55	_____	_____
15. 2,000.18	_____	_____

unit
6.9

Applications

1. A bicycle was sold for $69.99 in an area with a 7% retail sales tax. What was the tax on the bike?

ANSWERS

1. _____

2. a. _____

   b._____

3. _____

2. Goods with a retail price of $300 were sold in an area with a 5% retail tax. (a) What was the amount of the tax? (b) What was the total amount due from the customer?

3. If the 6% sales tax on a stove was $36, what was the price of the stove? (*Hint:* $P = \$36$, $R = 6\%$.)

4. After tax, a customer paid $525 for a TV set that had a $500 retail price. What was the rate of sales tax? [*Hint:* $500 = B, $25 ($525 − $500) = P.]

4. _____

5. _____

6. a. _____

   b. _____

5. Goods with a list price of $600 were sold at a 20% trade discount. What was the total amount due, including sales tax, if the goods were sold in an area with a 7% sales tax rate?

6. On a recent trip to the supermarket, David Wong bought goods that were priced as follows: $.69, $2.79, $12.99, $5.69, $4.99, and $2.99. If Bill shops in an area with a 7% retail sales tax, what is (a) the total amount of the purchase and (b) the amount of sales tax?

section

VII

Interest and Discounts

Computing Banker's Interest

After completing Unit 7.1, you will be able to:

1. Compute banker's interest.

The use of credit plays a very vital role in our nation's economy. Each business day countless numbers of businesses and individuals gain immediate possession of goods and services with the understanding that payment for these goods and services will be made in the future. Users of credit range from a high school student buying a $30 sweater on credit at a local department store to Donald Trump's $300,000,000 credit purchase of an airline.

≋ THE CHARGE FOR CREDIT

When borrowing money, or making a credit purchase, you will usually have to pay a fee for the use of credit. This fee is called *interest—it is money paid for the use of money*. The amount of interest you will be charged depends on three factors:

1. **Principal.** The principal is the amount borrowed.
2. **Rate.** The rate is the percent of interest being charged.
3. **Time.** The time is the length of time the debt will be outstanding.

To illustrate these factors, let's take an example. Bill Hardy wants to open his own carpet cleaning business. Bill estimates that he can get started in his new business with a $10,000 loan. So, Bill visited his local bank to talk with a loan officer. Bill found out that if he was willing to use the value of his house as security for the loan, the bank was willing to lend him the $10,000. Bill accepted these terms and agreed to repay the loan in one year, with interest at 10%.

The *principal* of Bill's loan is $10,000, the amount he borrowed. The *rate* of Bill's loan is 10%, the percent the bank charged. And the *time* of Bill's loan is one year, the length of time before Bill has to repay his loan.

INTEREST FOR A FULL YEAR

As stated earlier, interest is expressed as a rate (remember that "rate" means the same thing as "percent"). And, unless stated otherwise, the rate expressed is an *annual* rate. Thus, a rate of 10% means *10% of the principal for one year.* Therefore, to find the interest for one year, just multiply the principal by the rate.

Example

How much interest will Bill Hardy have to pay for the use of $10,000 for one year at 10%?

Solution

$$\frac{\text{Principal} \times \text{rate}}{\$10,000 \quad \times \quad .10} = \frac{\text{interest for one year}}{\textbf{\$1,000}}$$

THE INTEREST FORMULA

To find the interest for periods other than one year, we usually use the **interest formula,** which is:

$$\text{Interest} = \text{principal} \times \text{rate} \times \text{time} \qquad \text{or}$$

$$I \quad = \quad P \quad \times R \times T$$

To illustrate, let's find the interest on a $10,000 loan at 10% interest for 2 years:

$$\begin{array}{c} P \quad \times \ R \times T \\ \$10,000 \times .10 \times 2 = \textbf{\$2,000} \end{array}$$

Let's look at a few more examples.

Example 1

Find the interest on a $12,000 loan at 12% interest for 3 years.

Solution

$$\begin{array}{c} P \quad \times \ R \ \times T \\ \$12,000 \times .12 \times 3 = \textbf{\$4,320} \end{array}$$

Example 2

Find the interest on a $1,850 credit purchase of a sound system financed for 3 years at $12\frac{1}{2}$% interest.

Solution

$$\begin{array}{c} P \quad \times \ R \ \ \times T \\ \$1,850 \times .125 \times 3 = \textbf{\$693.75} \end{array}$$

Remember: $12\frac{1}{2}$% = 12.5% = 12.5% = .125

INTEREST FOR FRACTIONS OF A YEAR

Credit periods often will be for periods of time that are less than a year. For example, 30 days, 60 days, 90 days, 6 months, 9 months, and so on. When this happens, you state the T (time) in the interest formula as a fractional part of a year.

Time Expressed in Months

If the time of a loan is expressed in months, the number of months is placed over 12 (12 months in one year). So, 5 months becomes $\frac{5}{12}$; 3 months becomes $\frac{3}{12}$ or $\frac{1}{4}$; and 6 months becomes $\frac{6}{12}$ or $\frac{1}{2}$. To illustrate, let's compute the interest on a $1,200 loan at 12% interest for 3 months:

$$P \times R \times T$$
$$\$1,200 \times .12 \times \frac{1}{4}$$
$$= \$144 \times \frac{1}{4}$$
$$= \mathbf{\$36}$$

Remember: $\frac{3}{12} = \frac{1}{4}$

To solve this problem, we multiplied P by R ($1,200 × .12 = $144). Then we multiplied $144 by $\frac{1}{4}$ to get the interest charge. It is often easier to compute interest by using cancellation. For example, we can use cancellation to compute interest on the preceding problem as follows:

$$\frac{\overset{12}{\cancel{\$1,200}}}{1} \times \frac{\overset{3}{\cancel{12}}}{\underset{1}{\cancel{100}}} \times \frac{1}{\underset{1}{\cancel{4}}} = \$36$$

The rate is expressed as $\frac{12}{100}$ because percent means by the hundred.

Times Expressed in Days

To simplify the computation of interest when the time is expressed in days, the business community usually uses a 360-day year (called the **banker's year** or **ordinary interest**) instead of the actual year of 365 days. Use of the 365-day year will yield **exact interest** and is used by the U.S. government. We will study exact interest in Unit 7.2. Let's look at a couple of examples of interest calculation using a 360-day year.

Example 1

Find the interest on a $300, 90-day loan at a rate of 10%.

Solution

$$\frac{\overset{3}{\cancel{\$300}}}{1} \times \frac{\overset{5}{\cancel{10}}}{\underset{1}{\cancel{100}}} \times \frac{\overset{1}{\cancel{90}}}{\underset{\underset{2}{4}}{\cancel{360}}} = \$7.50$$

Example 2

Find the interest on an $800 loan for 60 days at 9% interest.

Solution

$$\frac{\overset{\overset{4}{\cancel{8}}}{\cancel{\$800}}}{1} \times \frac{\overset{3}{\cancel{9}}}{\underset{1}{\cancel{100}}} \times \frac{\overset{1}{\cancel{60}}}{\underset{\underset{\underset{1}{2}}{\cancel{6}}}{\cancel{360}}} = \$12$$

Self-Check

Tina borrowed $3,000 at 8% interest. What is her interest charge if the time of her loan was (a) 3 months; (b) 90 days?

Solution

(a) $60   (b) $60

Exercises

Find the interest for each of the following:

	Principal	Rate	Time	Interest
1.	$ 1,200	8%	1 year	$_____
2.	2,600	12%	3 years	_____
3.	200	14%	4 months	_____
4.	1,250	10%	9 months	_____
5.	4,500	11%	60 days	_____
6.	800	$12\frac{1}{2}$%	30 days	_____

	Principal	Rate	Time	Interest
7.	2,890	10%	45 days	_____
8.	12,000	15%	180 days	_____
9.	45,000	9%	120 days	_____
10.	62,500	12%	240 days	_____
11.	14,000	8%	90 days	_____
12.	95,000	14%	70 days	_____

Name _____ Date _____

Applications

1. What will be the interest charge on a $5,400 loan for 180 days at an interest rate of 9%?

ANSWERS

1. _____

2. _____

3. _____

2. Furgerson Wholesale Grocery Company needs to borrow $45,000 for 6 months. What will be the interest charge if a local bank is willing to make the loan at a rate of 12%?

3. To pay various bills, Don Myers borrowed $2,000 for 2 years at an interest rate of 10%. What will be his interest charge?

4. To take advantage of a special purchase price on word processors, Gary Lyons needs to borrow $3,560 for 30 days. What will be the interest charge if Gary can borrow the money at 9%?

ANSWERS

4. _____

5. _____

6. _____

5. Larry Swift is just starting a new law practice. The cost of law books, furniture, and equipment that he needs is $18,000. What will be the interest charge if Larry can borrow this amount for 2 years at 10% interest?

6. Harkon Building Company borrowed the following amounts during the last year:

$25,000 for 120 days at 12%
$38,000 for 2 months at 15%
$125,000 for 9 months at 11%

What was the total amount Harkon paid in interest during the year?

unit 7.1

Applications

1. What will be the interest charge on a $5,400 loan for 180 days at an interest rate of 9%?

ANSWERS

1. _____

2. _____

3. _____

2. Furgerson Wholesale Grocery Company needs to borrow $45,000 for 6 months. What will be the interest charge if a local bank is willing to make the loan at a rate of 12%?

3. To pay various bills, Don Myers borrowed $2,000 for 2 years at an interest rate of 10%. What will be his interest charge?

4. To take advantage of a special purchase price on word processors, Gary Lyons needs to borrow $3,560 for 30 days. What will be the interest charge if Gary can borrow the money at 9%?

ANSWERS

4. _____

5. _____

6. _____

5. Larry Swift is just starting a new law practice. The cost of law books, furniture, and equipment that he needs is $18,000. What will be the interest charge if Larry can borrow this amount for 2 years at 10% interest?

6. Harkon Building Company borrowed the following amounts during the last year:

$25,000 for 120 days at 12%
$38,000 for 2 months at 15%
$125,000 for 9 months at 11%

What was the total amount Harkon paid in interest during the year?

Computing Exact Interest and Solving for Unknowns

After completing unit 7.2, you will be able to:

1. Compute exact interest using a 365 year.
2. Compute exact interest using a calculator.
3. Use the interest formula to solve for unknowns.

In Unit 7.1, we learned that interest is the charge for credit. We also learned that most banks and private businesses use a 360-day year (the *banker's year*) as a base for computing interest. However, the U.S. government always uses a 365-day year to compute interest. Use of a 365-day year yields **exact interest,** which is more accurate than the banker's year because it is based on the true number of days in the year.

To compute exact interest, we still use the interest formula $I = P \times R \times T$, but we change the denominator of the time fraction from 360 to 365. Let's look at an example.

Example

Compute the exact interest due on a loan of $4,500 at 12% for 73 days.

Solution

$$\underbrace{\frac{\$4,500}{1} \times \frac{12}{100} \times \frac{73}{365}}_{P \quad \times \quad R \quad \times \quad T} = \$108$$

Remember: $12\% = .12 = \frac{12}{100}$

We canceled by 5

Self-Check

Find the exact interest on a loan of $900 at 10% for 30 days.

Solution

$$\frac{\overset{9}{\cancel{\$900}}}{1} \times \frac{10}{\underset{1}{\cancel{100}}} \times \frac{\overset{6}{\cancel{30}}}{\underset{73}{\cancel{365}}} = \frac{\$540}{73} = \$7.40$$

USING A CALCULATOR TO COMPUTE INTEREST

Computing exact interest is usually more time-consuming than banker's interest because 365 does not have as many factors as 360. So, when possible, using a calculator can save you time. To use a calculator, set up the problem like this:

$$\frac{\$4,500 \times 12\% \times 73}{365}$$

Now clear your calculator. Then follow these steps:

Step 1: Enter $4,500 and press the "×" key.

Step 2: Enter 12 and press the "%" key.

Step 3: Press the "×" key, enter 73, and press the "=" key.

Step 4: Press the "÷" key, enter 365, and press the "=" key.

After the last step, your calculator will display $108. Remember not to clear your calculator during the steps.

Self-Check

Using a calculator, find the interest on a loan of $1,500 at 12% for 90 days.

Solution

$$\frac{\$1,500 \times 12\% \times 90}{365} = \$44.38$$

FINDING UNKNOWNS USING THE INTEREST FORMULA

Thus far in our discussion of interest, we have used the interest formula ($I = P \times R \times T$) to find the amount of interest. Now we will use the interest formula to solve for principal, rate, and time. In all our calculations, we will use the banker's year of 360 days.

Finding the Principal

If a problem gives all factors except for the *principal,* use the following formula to find the principal:

$$\text{Principal} = \frac{\text{interest}}{\text{rate} \times \text{time}}$$

Example

Jan Reddi paid the bank $36 for a loan of 9% for 90 days. How much did she borrow?

Solution

Step 1: Set up the formula:

$$\text{Principal} = \frac{\text{interest}}{\text{rate} \times \text{time}} = \frac{\$36}{.09 \times 90/360}$$

Step 2: Multiply the denominator:

$$.09 \times \frac{90}{360} = (.09 \text{ times } 90 \text{ divided by } 360) = .0225$$

Step 3: Restate the formula:

$$\text{Principal} = \frac{\$36}{.0225} = \$1{,}600$$

Step 4: Check your work (using a calculator):

$$\frac{\$1{,}600 \times 9\% \times 90}{360} = \$36$$

Self-Check

What is the principal if the rate is 12%, the time is 45 days, and the interest is $18?

Solution

$$P = \frac{\$18}{.12 \times 45/360} = \frac{\$18}{.015} = \$1{,}200$$

Check:

$$\frac{\$1{,}200 \times 12\% \times 45}{360} = \$18$$

Finding the Rate

To find the *rate,* use this formula:

$$\text{Rate} = \frac{\text{interest}}{\text{principal} \times \text{time}}$$

Example

Jan Reddi paid the bank $36 for a $1,200 loan for 90 days. What was the interest rate?

Solution

Step 1: Set up the formula:

$$\text{Rate} = \frac{\text{interest}}{\text{principal} \times \text{time}} = \frac{\$36}{\$1,200 \times 90/360}$$

Step 2: Multiply the denominator:

$$\$1,200 \times \frac{90}{360} = \$300 \ (\$1,200 \text{ times } 90 \text{ divided by } 360)$$

Step 3: Restate the formula:

$$\text{Rate} = \frac{\$36}{\$300} = .12 = 12\%$$

Step 4: Check your work:

$$\frac{\$1,200 \times 12\% \times 90}{360} = \$36$$

Self-Check

What is the rate if the interest is $32.85, the principal is $1,825, and the time is 72 days?

Solution

$$\text{Rate} = \frac{\$32.85}{\$1,825 \times 72/360} = \frac{\$32.85}{\$365} = .09 = 9\%$$

Check:

$$\frac{\$1,875 \times 9\% \times 72}{360} = \$33.75$$

Finding the Time

To find the *time,* use this formula:

$$\text{Time (in years)} = \frac{\text{interest}}{\text{principal} \times \text{rate}}$$

Example

Jan Reddi paid the bank $36 for a $1,200 loan at 12%. How much time did she have to repay the loan?

Solution

Step 1: Set up the formula:

$$\text{Time} = \frac{\text{interest}}{\text{principal} \times \text{rate}} = \frac{\$36}{\$1,200 \times .12}$$

Step 2: Multiply the denominator:

$$\$1,200 \times .12 = \$144$$

Step 3: Restate the formula:

$$\text{Time} = \frac{\$36}{\$144} = .25 = 25\% \text{ of a year}$$

Step 4: Multiply the rate from Step 3 (.25) times 360 days to convert years to days:

$$360 \times .25 = 90 \text{ days}$$

Step 5: Check your work:

$$\frac{\$1,200 \times 12\% \times 90}{360} = \$36$$

Self-Check

What is the time if the interest is $76.80, the principal is $2,400, and the rate is 8%?

Solution

$$\text{Time} = \frac{\$76.80}{\$2,400 \times .08} = \frac{\$76.80}{\$192} = .40 = 40\% \times 360 = 144 \text{ days}$$

Check:

$$\frac{\$2,400 \times 8\% \times 144}{360} = \$76.80$$

Name _____ **Date** _____

Exercises

Find the exact interest for each of the following. Where necessary, round final answers correct to 2 places:

	Principal	Rate	Time	Interest
1.	$ 1,800	14%	90 days	$_____
2.	2,500	12%	40	_____
3.	1,095	9%	120	_____
4.	950	8%	73	_____
5.	1,400	10%	60	_____
6.	7,500	15%	125	_____

	Principal	Rate	Time	Interest
7.	9,800	12%	180	_____
8.	12,500	8%	92	_____
9.	25,800	10%	140	_____

Fill in the missing element in each of the following (assume a 360-day year in all calculations):

	Principal	Rate	Time	Interest
10.	$_____	12%	60 days	$31.00
11.	$4,500.00	_____	162 days	$202.50
12.	$9,000.00	9%	_____	$157.50

unit 7.2

Applications

1. Compute the exact interest on a loan for $12,200 for 75 days at 12% interest.

ANSWERS

1. _____

2. _____

3. _____

4. _____

2. Compute the banker's interest on an $18,000, 10%, 90-day loan. Then compute the exact interest. Which one is higher? Why?

3. Tim McCoy took out a student loan of $5,000 at 7% for 150 days. What is the exact interest?

4. Central Bank lent $250,000 for 250 days to one of its best customers. Interest was charged at a rate of 9%. Find the exact interest.

5. James Martin made an error when doing last year's taxes. He owes an additional $2,450. Find the total amount he must pay if he is assessed a 5% penalty and exact interest is charged on the total at 8% for 292 days.

5. _____

6. _____

7. _____

8. _____

6. Tom McPharland paid the bank $72 interest at 12% interest for 90 days? How much did he borrow?

7. What is the rate if the interest is $17.50, the principal is $2,100, and the time is 30 days?

8. Pat Farmer paid the bank $84 for a $6,000 loan at 7%. What is the time in days?

unit 7.3

Computing Interest Using Tables

After completing Unit 7.3, you will be able to:

1. Compute banker's interest using an interest table.
2. Compute exact interest using an interest table.

In Units 7.1 and 7.2, we learned to compute banker's and exact interest using the interest formula. We can also compute interest using tables. Table I on page 453 can be used to compute the banker's interest, and Table II on page 454 can be used to compute the exact interest. Both tables show the interest earned for $100 at various rates of interest and for different periods of time. To use either table, follow these steps:

Step 1: Find the rate of interest you need at the top of the table.

Step 2: Find the number of days you need in the "Time" column.

Step 3: Come down from the "Rate" column and across from the "Time" column until you meet. The point where you meet will be the amount of interest on $100 at that rate for the period of time. This amount is called the **interest factor.**

Step 4: Divide the principal by 100 (move the decimal point 2 places to the left). This will give you the number of $100s contained in the principal. This is the **principal factor.**

Step 5: Multiply the principal factor you obtained in Step 4 by the interest factor obtained in Step 3.

Example 1

Find the banker's interest on a principal of $650 for 25 days at $7\frac{1}{2}\%$ interest.

Solution

1. Use Table I and select the $7\frac{1}{2}\%$ column.
2. Find 25 days in the "Time" column.
3. Move down from $7\frac{1}{2}\%$ and across from 25 days until you meet. The interest factor is .5208.
4. Divide $650 by 100 and you get 6.5.
5. Multiply 6.5 by .5208. The interest is $3.39.

Example 2

Find the exact interest on a loan of $12,000 for 120 days at 8%.

Solution

Use Table II and select the 8% column. Move down from 8% and across from 120 days and you find an interest factor of 2.6301. Divide $12,000 by 100 and you get 120. Multiply 120 by 2.6301. The interest is $315.61.

If you need to compute interest for a rate (or time) that is not shown in the tables, you can combine rates or times. For example, what if you need to compute the banker's interest on a $800 loan for 90 days at 12% interest. As we can quickly see, 12% is not in the table, but 8% and 4% are. So, proceed as follows:

$$\$800 \div 100 = 8$$

Interest for 90 days at 8% = 8 × 2.0000   = $16.0000
Interest for 90 days at 4% = 8 × 1.0000   =   8.0000
Interest for 90 days at 12%   = $24.0000   = **$24.00**

You can likewise combine days, as the next example shows.

Example

Find the banker's interest on a loan of $4,800 for 72 days at 6% interest.

Solution

Table I does not have 72 days, but it does have 60 days and 12 days.

Interest factor for 60 days at 6% = $1.0000
Interest factor for 12 days at 6% =   0.2000
Interest factor for 72 days at 6% = $1.2000
$4,800 ÷ 100 = 48 × $1.2000   = $57.60 interest for 72 days at 6%

Table I

Time	4%	4½%	5%	5½%	6%	7%	7½%	8%
Banker's Interest ($100 For a 360-Day Year)								
1 day	.0111	.0125	.0139	.0153	.0167	.0194	.0208	.0222
2 days	.0222	.0250	.0278	.0306	.0333	.0389	.0417	.0444
3 days	.0333	.0375	.0417	.0458	.0500	.0583	.0625	.0667
4 days	.0444	.0500	.0556	.0611	.0667	.0778	.0833	.0889
5 days	.0556	.0625	.0694	.0764	.0833	.0972	.1042	.1111
6 days	.0667	.0750	.0833	.0917	.1000	.1167	.1250	.1333
7 days	.0778	.0875	.0972	.1069	.1167	.1361	.1458	.1556
8 days	.0889	.1000	.1111	.1222	.1333	.1556	.1667	.1778
9 days	.1000	.1125	.1250	.1375	.1500	.1750	.1875	.2000
10 days	.1111	.1250	.1389	.1528	.1667	.1944	.2083	.2222
11 days	.1222	.1375	.1528	.1681	.1833	.2139	.2292	.2444
12 days	.1333	.1500	.1667	.1833	.2000	.2333	.2500	.2667
13 days	.1444	.1625	.1806	.1986	.2167	.2528	.2708	.2889
14 days	.1556	.1750	.1944	.2139	.2333	.2722	.2917	.3111
15 days	.1667	.1875	.2083	.2292	.2500	.2917	.3125	.3333
16 days	.1778	.2000	.2222	.2444	.2667	.3111	.3333	.3556
17 days	.1889	.2125	.2361	.2597	.2833	.3306	.3542	.3778
18 days	.2000	.2250	.2500	.2750	.3000	.3500	.3750	.4000
19 days	.2111	.2375	.2639	.2903	.3167	.3694	.3958	.4222
20 days	.2222	.2500	.2778	.3056	.3333	.3889	.4167	.4444
21 days	.2333	.2625	.2917	.3208	.3500	.4083	.4375	.4667
22 days	.2444	.2750	.3056	.3361	.3667	.4278	.4583	.4889
23 days	.2556	.2875	.3194	.3514	.3833	.4472	.4792	.5111
24 days	.2667	.3000	.3333	.3667	.4000	.4667	.5000	.5333
25 days	.2778	.3125	.3472	.3819	.4167	.4861	.5208	.5556
26 days	.2889	.3250	.3611	.3972	.4333	.5056	.5417	.5778
27 days	.3000	.3375	.3750	.4125	.4500	.5250	.5625	.6000
28 days	.3111	.3500	.3889	.4278	.4667	.5444	.5833	.6222
29 days	.3222	.3625	.4028	.4431	.4833	.5639	.6042	.6444
30 days	.3333	.3750	.4167	.4583	.5000	.5833	.6250	.6667
60 days	.6667	.7500	.8333	.9167	1.0000	1.1667	1.2500	1.3333
90 days	1.0000	1.1250	1.2500	1.3750	1.5000	1.7500	1.8750	2.0000
120 days	1.3333	1.5000	1.6667	1.8333	2.0000	2.3333	2.5000	2.6667
150 days	1.6667	1.8750	2.0833	2.2917	2.5000	2.9167	3.1250	3.3333
180 days	2.0000	2.2500	2.5000	2.7500	3.0000	3.5000	3.7500	4.0000

100 x Rate x time = interest factor (i.e., interest on principal of $100.00)

Table II

Time	4%	$4\frac{1}{2}\%$	5%	$5\frac{1}{2}\%$	6%	7%	$7\frac{1}{2}\%$	8%

Exact Interest
($100 For a 365-Day Year)

Time	4%	$4\frac{1}{2}\%$	5%	$5\frac{1}{2}\%$	6%	7%	$7\frac{1}{2}\%$	8%
1 day	.0110	.0123	.0137	.0151	.0164	.0192	.0205	.0219
2 days	.0219	.0247	.0274	.0301	.0329	.0384	.0411	.0438
3 days	.0329	.0370	.0411	.0452	.0493	.0575	.0616	.0658
4 days	.0438	.0493	.0548	.0603	.0658	.0767	.0822	.0877
5 days	.0548	.0616	.0685	.0753	.0822	.0959	.1027	.1096
6 days	.0658	.0740	.0822	.0904	.0986	.1151	.1233	.1315
7 days	.0767	.0863	.0959	.1055	.1151	.1342	.1438	.1534
8 days	.0877	.0986	.1096	.1205	.1315	.1534	.1644	.1753
9 days	.0986	.1110	.1233	.1356	.1479	.1726	.1849	.1973
10 days	.1096	.1233	.1370	.1507	.1644	.1918	.2055	.2192
11 days	.1205	.1356	.1507	.1658	.1808	.2110	.2260	.2411
12 days	.1315	.1479	.1644	.1808	.1973	.2301	.2466	.2630
13 days	.1425	.1603	.1781	.1959	.2137	.2493	.2671	.2849
14 days	.1534	.1726	.1918	.2110	.2301	.2685	.2877	.3068
15 days	.1644	.1849	.2055	.2260	.2466	.2877	.3082	.3288
16 days	.1753	.1973	.2192	.2411	.2630	.3068	.3288	.3507
17 days	.1863	.2096	.2329	.2562	.2795	.3260	.3493	.3726
18 days	.1973	.2219	.2466	.2712	.2959	.3452	.3699	.3945
19 days	.2082	.2342	.2603	.2863	.3123	.3644	.3904	.4164
20 days	.2192	.2466	.2740	.3014	.3288	.3836	.4110	.4384
21 days	.2301	.2589	.2877	.3164	.3452	.4027	.4315	.4603
22 days	.2411	.2712	.3014	.3315	.3616	.4219	.4521	.4822
23 days	.2521	.2836	.3151	.3466	.3781	.4411	.4726	.5041
24 days	.2630	.2959	.3288	.3616	.3945	.4603	.4931	.5260
25 days	.2740	.3082	.3425	.3767	.4110	.4795	.5137	.5479
26 days	.2849	.3205	.3562	.3918	.4274	.4986	.5342	.5699
27 days	.2959	.3329	.3699	.4068	.4438	.5178	.5548	.5918
28 days	.3068	.3452	.3836	.4219	.4603	.5370	.5753	.6137
29 days	.3178	.3575	.3973	.4370	.4767	.5562	.5959	.6356
30 days	.3288	.3699	.4110	.4521	.4931	.5753	.6164	.6575
60 days	.6575	.7397	.8219	.9041	.9863	1.1507	1.2329	1.3151
90 days	.9863	1.1096	1.2329	1.3562	1.4794	1.7260	1.8493	1.9726
120 days	1.3151	1.4794	1.6438	1.8082	1.9726	2.3014	2.4657	2.6301
150 days	1.6438	1.8493	2.0548	2.2603	2.4657	2.8767	3.0822	3.2877
180 days	1.9726	2.2192	2.4657	2.7123	2.9589	3.4520	3.6986	3.9452

Self-Check

Jane Miller took out a loan for $12,000 for 30 days at 8% interest. Using the tables in this unit, find (a) the banker's interest and (b) the exact interest.

Solution

(a) $120 \times .6667 = \$80.00,$ **(b)** $120 \times .6575 = \$78.90$

Exercises

Use Table I to compute the interest for each of the following:

1. $1,500 at 8% for 30 days

2. $2,500 at $7\frac{1}{2}$% for 90 days

3. $12,000 at 12% for 60 days

4. $10,900 for 75 days at 10%

5. $125,600 for 200 days at 7%

6. $12,355 for 150 days at 16%

7. $1,890 for 60 days at $14\frac{1}{2}$%

8. $75,245.60 for 95 days at 8%

ANSWERS

1. _____
2. _____
3. _____
4. _____
5. _____
6. _____
7. _____
8. _____

Use Table II to compute interest for the following:

9. $1,800 for 75 days at 6%

10. $15,775 for 75 days at 8%

11. $245,600 for 120 days at $12\frac{1}{2}$%

12. $97,845 for 192 days at 8%

13. $14,245.18 for 90 days at 18%

14. $3,564.98 for 37 days at $7\frac{1}{2}$%

15. $3,000 for 300 days at 16%

9. _____

10. _____

11. _____

12. _____

13. _____

14. _____

15. _____

Name _____ Date_____

Applications

Use Tables I and II to solve the following problems:

1. Jane Trammel borrowed $12,000 for 90 days at an 8% banker's interest. (a) What is her interest charge? (b) What are her monthly payments if she repaid the loan over a 36-month period.

2. A loan for $12,000 was taken out for 180 days at a $7\frac{1}{2}$% interest. Find the interest using (a) the interest formula and (b) the banker's interest table.

3. The principal is $8,000, the rate is 8%, and the time is 92 days. (a) Compute the banker's interest by formula and table and (b) compute the exact interest by formula and table.

5. By how much does the banker's interest on a principal for $24,000 for 60 days at 12% differ from the exact interest on the same principal.

6. Use Table I to compute the banker's interest charge on a $4,000 loan for 77 days at 14%. Then use the PRT formula to check your answer.

ANSWERS

1. a. _____

   b._____

2. a. _____

   b._____

3. a. _____

   b._____

4. _____

5. _____

6. _____

unit 7.4

Computing Compound Interest

After completing Unit 7.4, you will be able to:

1. Compute compound interest for periods less than a year.
2. Compute compound interest for periods greater than a year.

In Unit 7.1, you learned that interest is the charge for credit. You also learned how to calculate interest using the formula $I = PRT$. To review this formula, let's assume you borrowed $1,000 at a rate of 12% for one year. Your interest charge would be:

$$P \quad \times \quad R \times T$$
$$\$1,000 \times .12 \times 1 = \$120$$

In this example, the $120 interest would be a charge to you because you had the use of someone else's money. But interest is a "two-way street." When you borrow money, you pay interest; when you save money with a savings bank (or other savings institution), you earn interest. Interest earned by you is usually "compounded," a topic we will discuss next.

〰 COMPUTING COMPOUND INTEREST

When you save with a savings bank, your money earns **compound interest.** This means that interest earned during each interest period is added to the principal before interest for the next period is calculated. Thus, the base on which interest is calculated becomes greater and greater.

To illustrate how compound interest works, let's assume you deposit $500 in a bank account that is currently paying 8% interest. And let's further assume you left your money on deposit with the bank for 2 years. Your compound interest for the 2 years is calculated as follows:

First year:

$$\$500 \times .08 \times 1 = \$40 \text{ interest}$$
$$\$500 + \$40 = \$540 \text{ compound amount}$$

Second Year:

$$\$540 \times .08 \times 1 = \$43.20 \text{ interest}$$
$$\$540 + \$43.20 = \$583.20 \text{ compound amount}$$

$583.20	compound amount
− 500.00	original principal
$ 83.20	compound interest

In this example, there were two interest periods, which gave us a compound interest calculation of $83.20. Compare this amount with interest calculated on the principal only (interest calculated on the principal only is called **simple interest**):

$$\$500 \times .08 \times 2 = \$80$$

As you can see, a simple interest calculation gives us $80 interest for 2 years, whereas compound interest results in interest of $83.20. So, your money would earn $3.20 more ($83.20 − $80.00) simply by the way the interest is calculated.

≈ COMPUTING COMPOUND INTEREST FOR SHORTER PERIODS

Interest is usually compounded more than once a year. For example, interest can be compounded monthly, quarterly (every 3 months), and semiannually (every 6 months). The more frequently the compounding, the more interest you earn—because the base grows faster. Let's see how much interest $500 would earn if interest at 8% is compounded semiannually (twice a year) for 2 years:

First interest period:

$$\$500 \quad \times .08 \times \frac{1}{2} \text{ year} = \$20 \text{ interest}$$
$$\underline{+ 20}$$
$$\$520 \quad \text{compound amount}$$

Second interest period:

$$\$520 \quad \times .08 \times \frac{1}{2} \text{ year} = \$20.80 \text{ interest}$$
$$\underline{+ 20.80}$$
$$\$540.80 \quad \text{compound amount}$$

Third interest period:

$$\$540.80 \quad \times .08 \times \frac{1}{2} \text{ year} = \$21.63 \text{ interest}$$
$$\underline{+ 21.63}$$
$$\$562.43 \quad \text{compound amount}$$

Fourth interest period:

$$\$562.43 \quad \times .08 \times \frac{1}{2} \text{ year} = \$22.50 \text{ interest}$$
$$\underline{+ 22.50}$$
$$\$584.93 \quad \text{compound amount}$$

Compound interest:

$$\begin{array}{rl} \$584.93 & \text{compound amount} \\ -\ \underline{500.00} & \text{original principal} \\ \$\ 84.93 & \text{compound interest} \end{array}$$

As you can see, when interest was compounded semiannually at 8% for 2 years, a $500 savings balance will earn $84.93. Compare this amount with the $83.20 figure we obtained earlier when interest on the same $500 was compounded annually for 2 years at 8%. By compounding the interest twice a year (instead of once a year), an additional $1.73 ($84.93 − $83.20) of interest was earned. Remember that the more frequently interest is compounded, the more interest you will earn.

USING A COMPOUND INTEREST TABLE

Compounding interest for a large number of interest periods can be very time-consuming. As a result, compound interest tables are often used.

Table 1 on page 462 shows compounded amounts for $1 at various interest rates for various time periods. To use the table, follow these steps:

Step 1: Place your finger on the correct number of interest periods in the far left column labeled "Number of Interest Periods."

Step 2: Find the appropriate interest rate at the top.

Step 3: Move across from the interest period and down from the interest rate until you meet; the number listed is the compound total for $1.

Step 4: Multiply the principal by the compound total for $1.

Example

Use Table 1 to calculate the compound interest on $600 for 5 years compounded annually at 8%.

Solution

Find period 5 in the first column and move across to the 8% column; you will find 1.46933. Multiply this number by the principal to find the compound amount:

$$\$600 \times 1.46933 = \$881.60 \text{ compound amount}$$

$$\begin{array}{rl} \$881.60 & \text{compound amount} \\ -\ \underline{600.00} & \text{principal} \\ \$281.60 & \text{compound interest} \end{array}$$

Table 1

Compound Interest Table (Amount of $1 at Compound Interest)

Number of Interest Periods	Rate per Period								
	1%	2%	3%	4%	5%	6%	8%	10%	12%
1	1.01000	1.02000	1.03000	1.04000	1.05000	1.06000	1.08000	1.10000	1.12000
2	1.02010	1.04040	1.06090	1.08160	1.10250	1.12360	1.16640	1.21000	1.25440
3	1.03030	1.06121	1.09273	1.12486	1.15763	1.19102	1.25971	1.33100	1.40493
4	1.04060	1.08243	1.12551	1.16986	1.21551	1.26248	1.36049	1.46410	1.57352
5	1.05101	1.10408	1.15927	1.21665	1.27628	1.33823	1.46933	1.61051	1.76234
6	1.06152	1.12616	1.19405	1.26532	1.34010	1.41852	1.58687	1.77156	1.97382
7	1.07214	1.14869	1.22987	1.31593	1.40710	1.50363	1.71382	1.94872	2.21068
8	1.08286	1.17166	1.26677	1.36857	1.47746	1.59385	1.85093	2.14359	2.47596
9	1.09369	1.19509	1.30477	1.42331	1.55133	1.68948	1.99900	2.35795	2.77308
10	1.10462	1.21899	1.34392	1.48024	1.62889	1.79085	2.15892	2.59374	3.10585
11	1.11567	1.24337	1.38423	1.53945	1.71034	1.89830	2.33164	2.85312	3.47855
12	1.12683	1.26824	1.42576	1.60103	1.79586	2.01220	2.51817	3.13843	3.89598
13	1.13809	1.29361	1.46853	1.66507	1.88565	2.13293	2.71962	3.45227	4.36349
14	1.14947	1.31948	1.51259	1.73168	1.97993	2.26090	2.93719	3.79750	4.88711
15	1.16097	1.34587	1.55797	1.80094	2.07893	2.39656	3.17217	4.17725	5.47357
16	1.17258	1.37279	1.60471	1.87298	2.18287	2.54035	3.42594	4.59497	6.13039
17	1.18430	1.40024	1.65285	1.94790	2.29202	2.69277	3.70002	5.05447	6.86604
18	1.19615	1.42825	1.70243	2.02582	2.40662	2.85434	3.99602	5.55992	7.68997
19	1.20811	1.45681	1.75351	2.10685	2.52695	3.02560	4.31570	6.11591	8.61276
20	1.22019	1.48595	1.80611	2.19112	2.65330	3.20714	4.66096	6.72750	9.64629
21	1.23239	1.51567	1.86029	2.27877	2.78596	3.39956	5.03383	7.40025	10.80385
22	1.24472	1.54598	1.91610	2.36992	2.92526	3.60354	5.43654	8.14027	12.10031
23	1.25716	1.57690	1.97359	2.46472	3.07152	3.81975	5.87146	8.95430	13.55235
24	1.26973	1.60844	2.03279	2.56330	3.22510	4.04893	6.34118	9.84973	15.17863
25	1.28243	1.64061	2.09378	2.66584	3.38635	4.29187	6.84848	10.83471	17.00006
26	1.29526	1.67342	2.15659	2.77247	3.55567	4.54938	7.39635	11.91818	19.04007
27	1.30821	1.70689	2.22129	2.88337	3.73346	4.82235	7.98806	13.10999	21.32488
28	1.32129	1.74102	2.28793	2.99870	3.92013	5.11169	8.62711	14.42099	23.88387
29	1.33450	1.77584	2.35657	3.11865	4.11614	5.41839	9.31727	15.86309	26.74993
30	1.34785	1.81136	2.42726	3.24340	4.32194	5.74349	10.06266	17.44940	29.95992

USING A TABLE FOR PERIODS OF LESS THAN A YEAR

In the preceding examples, we used Table 1 to calculate compound interest on an annual basis. When interest is compounded annually, the number of interest periods is the same as the number of years. But, when using Table 1 to compound interest for shorter periods, you must adjust the interest periods and the annual rate. To do so, follow these steps:

Step 1: Multiply the number of years by the number of times interest will be compounded in the year.

Step 2: Divide the interest rate by the number of times interest will be compounded during the year.

Step 3: Using these adjusted figures from Steps 1 and 2, calculate the compound interest.

Example

Use Table 1 to calculate the compound interest on $300 for 3 years compounded quarterly at 8%.

Solution

Quarterly compounding = 4 times a year

Step 1: Multiply 3 years by 4:

$$3 \text{ years} \times 4 = 12 \text{ interest periods}$$

Step 2: Divide the annual interest rate by 4:

$$8\% \div 4 = 2\%$$

Step 3: Find period 12 in the first column of Table 1 and move across on the same line to the 2% column; you will find 1.26824. Multiply this figure by the principal to get the compound amount:

$$\$300 \times 1.26824 = \$380.47 \text{ compound amount}$$

$$
\begin{array}{rl}
\$380.47 & \text{compound amount} \\
-\ \underline{300.00} & \text{principal} \\
\$\ 80.47 & \textbf{compound interest}
\end{array}
$$

Self-Check

Find the amount of compound interest earned on a $3,000 investment at 12% for 2 years if interest is compounded monthly.

Solution

Monthly compounding = 12 times a year:

Step 1: 2 years × 12 = 24 interest periods

Step 2: 12% ÷ 12 = 1%

Step 3: Find 24 in the first column of Table 1 and move across to the 1% column; you will find 1.26973. Multiply this figure by the principal:

$3,000 × 1.26973 = $3,809.19 compound amount
$3,809.19 − $3,000 = **$809.19 compound interest**

unit 7.4 Exercises

Without using Table 1, find the compound interest for each of the following:

	Principal	Rate	Time	Compounding Period	Compound Interest
1.	$ 800	12%	1 year	quarterly	$_____
2.	1,200	8%	3 years	annually	_____
3.	5,000	10%	5 years	semiannually	_____
4.	600	9%	2 years	quarterly	_____

Use Table 1 to complete the following:

	Principal	Rate	Time	Compounding Period	Compound Interest
5.	$ 4,500	6%	3 years	annually	$_____
6.	12,000	10%	5 years	semiannually	_____
7.	6,000	8%	4 years	quarterly	_____
8.	20,000	12%	2 years	monthly	_____

Name _____ **Date** _____

Applications

1. Lyle Burgman deposited $700 in his savings account 2 years ago. His bank is paying 8% interest compounded quarterly. How much interest has he earned?

1. _____

2. _____

2. Julie Gamble deposited $2,000 in her savings account 5 years ago. What is the balance of her account today if she earned 12% interest compounded quarterly?

3. What interest would a $12,000 savings balance earn in one year if the bank is paying 7% interest compounded semiannually?

3. _____

4. _____

4. How much interest would $6,000 earn in 2 years if it is deposited in an account paying 12% interest compounded monthly?

Installment Purchases and Computing the True Rate of Interest

After completing Unit 7.5, you will be able to:

1. Compute the cost of buying on time.
2. Determine the real rate of interest on a loan.

Buying goods by making a down payment and paying the balance over a period of time is called **installment purchasing,** or **buying on time.** Installment purchases play a huge role in our nation's economy, as most of us make installment purchases on a regular basis.

Although some merchants will allow you to make installment payments without an interest charge, most require interest based on the amount of time it takes to pay off the purchase. Also, the purchaser is generally charged various fees, such as an application fee for the loan, credit report fees, and perhaps credit life insurance. Credit life insurance guarantees that if anything happens to the purchaser, the debt will be paid off by an insurance company, rather than by the family of the purchaser. These additional costs are called the **added cost** of borrowing.

To find the added cost of an installment purchase, follow these steps:

Step 1: Multiply the amount of each payment by the number of payments.

Step 2: Add to the total in Step 1 the amount of the down payment plus any other up-front fee that was not financed along with the product. This gives you the total installment price.

Step 3: Subtract the cash purchase price of the product from the total you obtained in Step 2.

Example

Sterling James purchased a large screen TV for $2,100 on the installment plan. He made a down payment of $300 and will pay off the balance in 18 monthly installments of $118.00. What is the added cost of buying on time?

Solution

$$\begin{array}{rl}
\$2,124.00 & \text{total of monthly payments } (\$118.00 \times 18) \\
+\ \ \ \ 300.00 & \text{down payment} \\
\hline
\$2,424.00 & \text{total installment price} \\
-\ \ 2,100.00 & \text{cash price} \\
\hline
\$\ \ \ 324.00 & \text{added cost of buying on time}
\end{array}$$

Self-Check

The cash price of a VCR is $300. What is the added cost of buying on time if it is purchased on the installment plan with a $50 down payment and 12 equal payments of $26.50, and a $15 credit application fee is paid upfront.

Solution

$$\begin{array}{rl}
\$318 & (\$26.50 \times 12) \\
+\ \ 65 & \text{down payment and credit application fee} \\
\hline
\$383 & \text{total installment price} \\
-\ \ 300 & \text{cash price} \\
\hline
\$\ \ 83 & \text{added cost of buying on time}
\end{array}$$

FINDING THE REAL ANNUAL RATE OF INTEREST

When personal, family, or household loans do not exceed $25,000, the Truth-in Lending Act requires the lender to state **the true annual rate of interest,** which is the actual rate of interest that you are paying for buying on credit or taking out a loan. If you know the monthly rate that you are being charged, it is simple to determine the annual rate: Multiply the monthly rate by 12. For example, if a loan specifies interest at a rate of $1\frac{1}{2}\%$ a month, the annual rate is 18% ($1\frac{1}{2}\% \times 12$ months).

If the monthly rate is not given, you can use the **constant ratio formula,** which is expressed as follows:

$$R = \frac{24 \times I}{P \times (n + 1)}$$

where R = the true annual rate of interest
  I = the interest charge
  P = the principal of the loan (or the price of the product purchased)
  n = the number of payments

Does not incl the down payment: you don't borrow that

Example

Given the following factors, what is the true annual rate of interest?

$$P = \$300$$
$$I = \$90$$
$$n = 23$$

Solution

$$R = \frac{24 \times 90}{300 \times (23 + 1)} = \frac{\overset{1}{\cancel{24}} \times \overset{3}{\cancel{90}}}{\underset{10}{\cancel{300}} \times \underset{1}{\cancel{24}}} = \frac{3}{10} = .30 \ (30\%)$$

Self-Check

Bill Willis's $500 loan is to be repaid in nine equal installments of $65 each. Find the true annual rate of interest.

Solution

$65 \times 9 = \$585$ total payments; $585 - \$500 = \$85$ interest charge

$$R = \frac{24 \times 85}{500 \times (9 + 1)} = \frac{\overset{6}{\underset{\cancel{12}}{\cancel{24}}} \times \overset{17}{\cancel{85}}}{\underset{\underset{50}{\cancel{100}}}{\cancel{500}} \times \underset{5}{\cancel{10}}} = \frac{102}{250} = .408 = 40.8\%$$

Cancellations:

$$500 \div 5 = 100$$
$$85 \div 5 = 17$$
$$24 \div 2 = 12$$
$$10 \div 2 = 5$$
$$100 \div 2 = 50$$
$$12 \div 2 = 6$$

Name _____ **Date** _____

Exercises

1. Complete the following table:

Product	Down Payment	Number of Payments	Monthly Payments	Installment Price	Cash Price	Cost of Buying on Time
Microcomputer	None	24	$48.50	$_____	$995.00	$_____
Laser printer	$100	12	28.00	_____	349.99	_____
Stereo	200	18	30.45	_____	549.95	_____
Microwave	None	15	18.50	_____	229.99	_____

2. Find the annual rate of interest from the following monthly rates:

Monthly Rate	Annual Rate
a. 1.6%	_____%
b. 1.5	_____
c. .8	_____
d. 1.44	_____

3. Find the true annual rate of interest in each of the following problems:

	Down Payment	Number of Payments	Monthly Payments	Cash Price	True Annual Rate
a.	$100.00	15	$30.00	$500.00	_____%
b.	20.00	15	32.00	400.00	_____
c.	50.00	8	50.00	395.00	_____
d.	75.00	18	25.55	450.00	_____

Name _____ Date_____

Applications

1. Wendy Cyrus purchased a $2,000 oak desk on the installment plan for her law office. She made a down payment of $200 and paid 12 monthly payments of $175. What was the added cost of buying on time?

2. Jorge White purchased a used car for $9,200. After a down payment of 10%, he made 36 payments of $247.88. How much more did he pay to buy the car on time?

3. Valley Produce Company purchased a new electronic cash register for $2,900. The company paid 15% down and financed the difference for 18 months at $152.08 per month. How much did it cost to finance the purchase?

4. Bi-City Flower Shop bought a new delivery van for $18,000. After paying $2,000 down, the monthly payments were $348 for 60 months. What was the total financing charge?

5. Lisa Love's credit card company charges her a monthly rate of 1.75%. What is her annual rate?

6. Ben Reed's loan of $550 is to be repaid in 40 weekly installments of $15. What is his true annual rate of interest?

7. Cynthia Glenn's loan of $1,500 is to be repaid over 24 months at a rate of $\frac{7}{10}$% a month. Find the total finance charge on this loan.

unit 7.6

Computing Interest on Unpaid Balances

After completing Unit 7.6, you will be able to:

1. Compute interest on the unpaid balance of an installment purchase.

When buying on time, you usually make a down payment and pay the balance, with interest, over a period of time. Since each payment you make—starting with the very first one—includes part of the interest charge, the fairest basis on which to compute the interest is on the unpaid balance. The interest charge on the unpaid balance is computed by this formula:

$$I = \frac{\text{Interest for first month} \times \text{number of months} + 1}{2}$$

Example

A lawn tractor was purchased on May 1 for $1,200 less a down payment of $200. The $1,000 balance is to be repaid in equal monthly installments over a period of 4 months with interest on the unpaid balance at 9%.

a. Find the interest charge based on the unpaid balance.
b. Find the amount of each payment.

Solution

a. $I = \dfrac{\text{Interest for first month} \times \text{number of months} + 1}{2} = \dfrac{\$7.50^* \,(4 + 1)}{2} = \dfrac{\$37.50}{2} = \$18.75$

b. $\$1{,}000 + \$18.75 = \dfrac{\$1018.75}{4} = \$254.69$

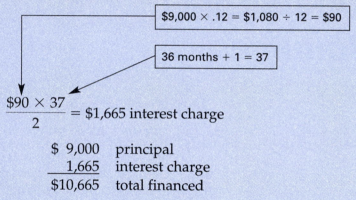

Self-Check

A car costing $9,000 is financed for 36 months at 12% interest. What are the monthly payments if interest is computed on the unpaid balance?

Solution

$\$9{,}000 \times .12 = \$1{,}080 \div 12 = \$90$

$36 \text{ months} + 1 = 37$

$\dfrac{\$90 \times 37}{2} = \$1{,}665 \text{ interest charge}$

$\begin{array}{rl} \$\;9{,}000 & \text{principal} \\ \underline{1{,}665} & \text{interest charge} \\ \$10{,}665 & \text{total financed} \end{array}$

$\dfrac{\$10{,}665}{36 \text{ months}} \text{ months} = \$296.25 \text{ monthly payments}$

$^*\$1{,}000 \times .09 = \dfrac{\$90}{12} = \$7.50$

Name _____ Date _____

Exercises

Assuming interest is computed on the unpaid balance, complete the
following chart:

	Principal	Interest Rate	Time	Total Interest Charge	Monthly Payment
1.	$12,400	14%	24 months	$_____	$_____
2.	18,400	12%	60	_____	_____
3.	1,200	10%	12	_____	_____
4.	2,800	9%	8	_____	_____
5.	695	18%	15	_____	_____

Name _____ Date _____

Applications

1. John Henzel purchased a complete stereo system for $6,000, plus sales taxes of 7%. He paid 10% down and financed the balance at 12% on the unpaid balance for 24 months. What was (a) his total finance charge and (b) the amount of each payment.

ANSWERS

1. a. _____

   b. _____

2. _____

3. _____

2. Refer to problem 6. Compute the amount of John's monthly payment if he had been charged simple interest based on the $I = PRT$ formula. Why are his payments more?

3. A new car costing $18,000 is financed for 60 months at 10% on the unpaid balance. What is the monthly payment?

unit 7.7

Promissory Notes

After completing Unit 7.7, you will be able to:

1. Describe the parts of a promissory note.
2. Find the due date of a note.
3. Find the maturity value of a note.
4. Find the proceeds from a discounted note.

If you borrowed money from a friend, your friend might ask you to sign an IOU ("I Owe You"), which is a written promise to repay an amount of money at some future time. When borrowing money from a bank (or other lending institution), you usually sign a **promissory note,** which is a formal IOU. A typical promissory note follows, with a list of special terms:

1 June 12, 20 XX

2 Sixty (60) days _____ after date **9** I promise to pay to

the order of **3** Alli Sterling _____ **4** $1,200.00

One thousand two hundred and no/100 _____ Dollars

5
Payable at _____ Interstate National Bank _____

6
(Signed) _____ Bill Tucker _____

7
Value received, with interest at __ 9% __

8
DUE _____ Aug. 11, 20XX _____

1. The **date** the note was signed was June 12, 20XX.
2. The **time,** or *term,* of the note is 60 days.
3. The **payee** of the note is Alli Sterling. Alli is the person who will be paid when the note is due.
4. The **principal,** or *face value,* of the note is $1,200.00. The principal, which is written in both figures and words, is the amount borrowed or the amount of credit received.
5. The note is payable at the Interstate National Bank where the payee most likely has an account.
6. The **maker** of the note is Bill Tucker. Bill is the person who will pay Alli when the note is due.
7. The **rate** of interest charged on the note is 9%. Some notes do not bear interest, in which case they are referred to as *noninterest bearing.*
8. The **due date** of the note is August 11, 20XX. This is the date Bill will make payment to Alli. The due date is also called the **maturity date.**

 ## FINDING THE DUE DATE OF A NOTE

The time of a note is usually expressed in either months or days. If the time is in months, the note is due on the same day of the month, so many months later. For example, a one-month note dated March 23 is due April 23. Likewise, a three-month note dated July 26 is due October 26. A one-month note dated March 31 presents a tiny problem, however, because there is no April 31. When this happens, the last day of the month becomes the due date. Thus, a one-month note dated March 31 note would be due on April 30.

When the note is expressed in days, you will need to count the exact days. One way to do this is to take a calendar and count ahead the appropriate number of days. This procedure, however, is time-consuming, and you may not always have a calendar handy. There is another way to determine the due date without the use of a calendar. For example, refer to the note Bill Tucker gave to Alli Sterling (shown earlier). The August 11 due date can be determined by deducting the date of the note (June 12) from the 30 days in June, and by continuing to count days until you reach 60 days:

Number of days remaining in June (30 − 12)	18
Days in July	31
Total days at the end of July	49
Days in August needed to reach 60 (60 − 49)	11 ◄——due date (August 11)
Term of note	60 days

Similarly, the due date of a 90-day note dated August 5 is determined as follows:

Number of days remaining in August (31 − 5)	26
Days in September	30
Days in October	31
Total days at the end of October	87
Days in November needed to reach 90	3 ◄——due date (November 3)
Term of note	90 days

A third method uses a table of the numbers of the days of the year, as follows:

Number of the Days of the Year

Day of Month	Jan.	Feb.	Mar.	Apr.	May	June	July	Aug.	Sept.	Oct.	Nov.	Dec.
1	1	32	60	91	121	152	182	213	244	274	305	335
2	2	33	61	92	122	153	183	214	245	275	306	336
3	3	34	62	93	123	154	184	215	246	276	307	337
4	4	35	63	94	124	155	185	216	247	277	308	338
5	5	36	64	95	125	156	186	217	248	278	309	339
6	6	37	65	96	126	157	187	218	249	279	310	340
7	7	38	66	97	127	158	188	219	250	280	311	341
8	8	39	67	98	128	159	189	220	251	281	312	342
9	9	40	68	99	129	160	190	221	252	282	313	343
10	10	41	69	100	130	161	191	222	253	283	314	344
11	11	42	70	101	131	162	192	223	254	284	315	345
12	12	43	71	102	132	163	193	224	255	285	316	346
13	13	44	72	103	133	164	194	225	256	286	317	347
14	14	45	73	104	134	165	195	226	257	287	318	348
15	15	46	74	105	135	166	196	227	258	288	319	349
16	16	47	75	106	136	167	197	228	259	289	320	350
17	17	48	76	107	137	168	198	229	260	290	321	351
18	18	49	77	108	138	169	199	230	261	291	322	352
19	19	50	78	109	139	170	200	231	262	292	323	353
20	20	51	79	110	140	171	201	232	263	293	324	354
21	21	52	80	111	141	172	202	233	264	294	325	355
22	22	53	81	112	142	173	203	234	265	295	326	356
23	23	54	82	113	143	174	204	235	266	296	327	357
24	24	55	83	114	144	175	205	236	267	297	328	358
25	25	56	84	115	145	176	206	237	268	298	329	359
26	26	57	85	116	146	177	207	238	269	299	330	360
27	27	58	86	117	147	178	208	239	270	300	331	361
28	28	59	87	118	148	179	209	240	271	301	332	362
29	29	—	88	119	149	180	210	241	272	302	333	363
30	30	—	89	120	150	181	211	242	273	303	334	364
31	31	—	90	—	151	—	212	243	—	304	—	365

To illustrate, refer again to the note Bill Tucker gave to Alli Sterling. First look at the table for the number for June 12, and you find 163. This means that June 12 is the 163rd day of the year. Next, add the time of the note (60 days) to 163 to get 223. Now look up 223 in the table. You will find that August 11 is the due date.

Try a note dated January 11 with a time of 90 days. January 11 is day 11. Add 90 to 11 and you get 101. Day 101 is April 11.

FIND THE MATURITY VALUE OF A NOTE

The **maturity value** of a note is the amount that must be repaid when the note is due. It is computed as follows:

$$\text{Principal} + \text{interest} = \text{maturity value}$$

To illustrate, the 60-day note Bill Tucker gave to Alli Sterling (illustrated earlier) has a principal of $1,200 and a rate of 9%. We can find the maturity value as follows:

Step 1: Find the interest:

$$\frac{\$1,200 \times 9\% \times 60}{360} = \$18$$

Step 2: Add the interest to the principal:

$$\frac{\text{Principal}}{\$1,200} + \frac{\text{Interest}}{\$18} = \frac{\text{Maturity Value}}{\$1,218}$$

As we can see, the maturity value of this note is $1,218. This is the amount that Bill must pay Alli when the note reaches its due date (August 11, 20XX). By the way, if a note is noninterest bearing, the maturity value and the principal will be the same.

≋ DISCOUNTING A PROMISSORY NOTE

An important feature of a note is that it is **negotiable;** that is, you can transfer it to another party by **endorsement** (signature). To illustrate, let's return once again to the note illustrated at the beginning of this unit. By looking at the note, we can clearly see that Bill must pay Alli the maturity value of the note on August 11. But, what if Alli needs money before August 11? What if she needs her money by July 22? She has the option of selling the note at a discount to another party (usually another person or a loan company). Alli would receive her money—less the discount—on July 22, and the party who bought the note would hold it to August 11 and collect the full maturity value. Let's assume Alli sold the note to Frank Winners on July 22. Let's further assume Frank charged a discount rate of 12%. The steps in the process follow:

Step 1: Determine the maturity value of the note. Earlier we determined the maturity value of the note as follows:

$$\frac{Principal \; + \; Interest}{\$1,200 \quad + \quad \$18} = \frac{Maturity \; Value}{\$1,218}$$

Step 2: Determine the due date of the note. Earlier we determined the due date to be August 11, 20XX:

Step 3: Determine the discount period.

The **discount period** is the number of days from the date that the note is discounted until the due date. Alli discounted the note on July 22, and the note is due on August 11. The number of days between these dates is the discount period. We find those days as follows:

Days in July	31
Less date the note was discounted	22
Days remaining in July	9
Days in August needed to reach due date	11
Discount period	20 days

Step 4: Compute the amount of discount, as follows:

$$\frac{Maturity \; Value}{\$1,218} \times \frac{Rate \; of \; Discount}{12\%} \times \frac{Discount \; Period}{20/360} = \frac{Discount}{\$8.12}$$

Step 5: Compute the proceeds of the note. The **proceeds** is the amount of money Alli will receive, determined as follows:

Maturity value	$1,218.00
Less discount	− 8.12
Proceeds	$1,209.88

Alli receives $1,209.88 on July 22. Had she waited until August 11, she would have received the full maturity value of $1,218. She was willing to receive the lesser amount in order to get the use of the money sooner. Frank Winners will hold the note until August 11 and present it to Bill Tucker for payment. Bill will then pay Frank the maturity value of $1,218.

Exercises

Name _____ Date_____

1. Study this note and answer the questions that follow:

_____ May, 18 20 __XX__

_____ Ninety (90) days _____ after date __9__ promise to pay to

the order of _____ Larry Best _____ $1,800.00 _____

One thousand eight hundred and no/100 _____ Dollars

Payable at __ Thompson National Bank _____

(Signed) ___ Lowell Cummins _____

Value received, with interest at __10%___ DUE _____

a. What is the date of the note?

b. What is the time of the note?

c. Who is the maker of the note?

d. What is the principal?

e. Where is the note payable?

f. Who is the payee of the note?

g. Is the note interest bearing or noninterest bearing?

h. What is the due date of the note?

i. What is the maturity value of the note?

ANSWERS

1. a. _____

b. _____

c. _____

d. _____

e. _____

f. _____

g. _____

h. _____

i. _____

2. Determine the due date of the following notes:

Note	Date	Time	Due Date
A	July 2, 20X1	30 days	_____
B	August 15, 20X1	90 days	_____
C	November 20, 20X1	120 days	_____
D	March 23, 20X2	1 month	_____
E	July 26, 20X3	3 months	_____
F	October 31, 20X1	1 month	_____

2. _____

3. _____

4. _____

3. Determine the maturity value of the following notes:

Note	Principal	Rate	Time	Maturity Value
A	$2,450	12%	90 days	$_____
B	1,560	10%	120 days	_____
C	3,600	9%	3 months	_____
D	1,800	8%	2 years	_____

4. For each of the following notes, compute the proceeds:

Note	Date of Note	Time of Note	Principal	Interest Rate	Rate of Discount	Date of Discount	Proceeds
A	July 11	30 days	$1,200	10%	12%	July 21	$_____
B	June 10	60 days	2,000	9%	10%	June 30	_____
C	March 5	45 days	6,000	12%	12%	March 20	_____
D	June 8	2 months	2,400	8%	9%	June 29	_____

Applications

Name _____ Date _____

1. On August 1, 20XX, Ben Taylor borrowed $5,000 by issuing a 90-day, 12% note. What is (a) the due date of the note and (b) the maturity value of the note.

ANSWERS

1. a. _____

   b. _____

2. _____

3. a. _____

   b. _____

2. Benson Company receives a $540, 120-day non-interest-bearing note from Bill Phillips on June 7, 20XX. The note is discounted at 8% on August 22, 20XX. Find the proceeds.

3. A 7%, 90-day note for $2,400 is discounted for 60 days at 10%. Find (a) the discount and (b) the proceeds.

4. Li Company issued a 4-month, $900, 12% note to Mitchell Company on May 24, 20X1. Mitchell Company discounted the note on June 26, 20X1, at 10%. Find (a) the amount of discount and (b) the proceeds of the note.

5. Find the proceeds of King Company's $900 2-month note dated March 25 if it is discounted on April 10 (a) at 10% and (b) at 12%.

unit 7.8

Trade Discounts

After completing Unit 7.8, you will be able to:

1. Compute trade discounts.

Producers and distributors of products often issue catalogs that show the **list price** or the **catalog price** of the merchandise they sell. A wholesale auto parts distributor, for example, could issue a catalog that describes the company's product line and gives a price for each product. These catalogs would then be distributed to the wholesaler's customers (mostly retail auto parts stores).

We can thus conclude that a seller's catalog serves two main purposes: (1) it gives a description of goods that are available and (2) it gives the current prices of the goods. But, what happens when the price of an item (or items) changes? Does this mean that the seller must pay to reprint the catalog so that the new prices will be shown in the catalog?

Usually it is too expensive to reprint a catalog each time the price of merchandise changes. Instead, many producers and distributors issue, in addition to catalogs, sheets of **trade discounts** that may be available. A trade discount is a percentage reduction in the list price of merchandise. When the price of merchandise changes, the seller can update the price by simply updating the discount sheet. For example, an item can be listed in a catalog for, say, $500 less a 10% discount. If the price of the item goes up, the 10% discount can be lowered, or dropped altogether.

Trade discounts also allow the seller greater flexibility when selling to different classes of customers, or when selling larger amounts. For example, sellers often offer a larger discount to customers who buy in large quantities.

The amount of a trade discount is determined as follows:

$$\begin{array}{r} \text{List price of merchandise} \\ \times\ \underline{\text{trade discount rate}} \\ =\ \text{amount of trade discount} \end{array}$$

The amount of trade discount is then subtracted from the list price to get the **net price** (**list price** − **amount of trade discount** = **net price**). The net price is the amount the customer actually pays for the goods.

Example 1

An item is listed at $800 less a 20% trade discount. Find the amount of the trade discount and the net price.

Solution

Step 1: Calculate the amount of trade discount:

$$\begin{array}{rl} \$800 & \text{list price} \\ \times \ \underline{\ \ .20} & \text{trade discount rate} \\ \$160 & \text{amount of trade discount} \end{array}$$

Step 2: Calculate the net price:

$$\begin{array}{rl} \$800 & \text{list price} \\ -\ \underline{\ 160} & \text{amount of trade discount} \\ \$640 & \text{net price} \end{array}$$

Example 2

What is the trade discount and net price of an item listed at $1,200 with a 40% trade discount?

Solution

$$\begin{array}{rl} \$\ 1{,}200 & \text{list price} \\ \times \underline{\ \ \ \ \ .40} & \text{trade discount rate} \\ \$480.00 & \text{amount of trade discount} \end{array}$$

$$\begin{array}{rl} \$1{,}200 & \text{list price} \\ -\ \underline{\ \ \ 480} & \text{amount of trade discount} \\ \$\ 720 & \text{net price} \end{array}$$

When calculating the amount of trade discount, we are actually using the formula $B \times R = P$:

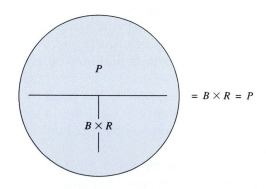

$$= B \times R = P$$

B = List price of merchandise

R = Trade discount rate

P = Amount of trade discount

Example

What is the trade discount on goods listed at $600 less 40%?

Solution

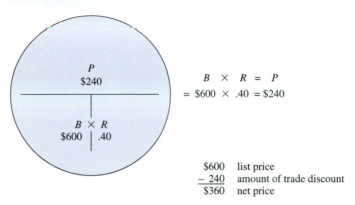

$$B \times R = P$$
$$= \$600 \times .40 = \$240$$

$$
\begin{array}{ll}
\$600 & \text{list price} \\
-\ 240 & \text{amount of trade discount} \\
\hline
\$360 & \text{net price}
\end{array}
$$

Helpful Hint: Another way to find the net price is to multiply the list price of merchandise by the **complement** of the trade discount rate. To get the complement of a rate, you subtract the rate from 100%. In the preceding example, for instance, the complement of 40% is 60% (100% − 40%). You can then find the net price as follows:

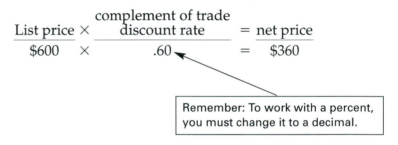

If you need to find the amount of trade discount, you can then subtract the net price from the list price:

$$
\begin{array}{ll}
\$600 & \text{list price} \\
-\ 360 & \text{net price} \\
\hline
\$240 & \text{amount of trade discount}
\end{array}
$$

Self-Check

Merchandise costing $2,000 was sold at a 30% trade discount. Find (a) the amount of discount and (b) the net price.

Solution

(a) $2,000 × .30 = $600 discount
(b) $2,000 − $600 = $1,400 net price

Work Space

Name _____ Date_____

Exercises

Find the trade discount and net amount for each of the following:

	List Price	Rate of Trade Discount	Amount of Trade Discount	Net Price
1.	$ 500	5%	$_____	$_____
2.	850	12%	_____	_____
3.	1,800	$10\frac{1}{2}\%$	_____	_____
4.	2,540	25%	_____	_____
5.	9,000	$33\frac{1}{3}\%$	_____	_____
6.	8,400	30%	_____	_____
7.	6,450	42%	_____	_____
8.	2,459	$16\frac{2}{3}\%$	_____	_____

9. 4,200   28%   _____   _____

10. 9,560   31%   _____   _____

11. 6,280 $8\frac{1}{3}\%$ _____   _____

12. 7,550   35%   _____   _____

13. 8,490   22%   _____   _____

14. 8,600   17%   _____   _____

15. 6,240   27%   _____   _____

16. 5,400 $37\frac{1}{2}\%$ _____   _____

Find the complement of the following rates:

17. 15%   18. 25%   19. 42%

20. 10%   21. 24%   22. 32%

23. 45%   24. 8%   25. 21%

17. _____

18. _____

19. _____

20. _____

21. _____

22. _____

23. _____

24. _____

25. _____

Applications

1. What is the net price of a personal computer listed at $1,200 with a trade discount of 20%?

2. Ledbetter Company bought a drill press that was listed at $48,500 with a trade discount of 25%. (a) What is the amount of trade discount? (b) What is the net price of the item?

3. Jim Swazey is thinking of buying a new air-conditioning unit for his office. He found one with a list price of $2,800 and no discount. He found another with a list price of $3,400 and a trade discount of 12%. Which should he buy? Why?

4. Doyle Lowell bought a computer keyboard and got a $50 trade discount. If the keyboard had a list price of $300, what rate of discount did Doyle get? (*Hint:* $B = \$300$, $P = \$50$, find R.)

ANSWERS

1. _____

2. a. _____

   b. _____

3. _____

4. _____

5. What is the net price of a grandfather clock that has a list price of $1,800 and a trade discount of $33\frac{1}{3}$%?

6. Carlo Rojas paid $1,200 for an item that was listed for $1,800. What rate of discount did he get? (*Hint:* $1,200 = the complement of the trade discount.)

7. Campus Bookstore paid $7,500 net for textbooks this semester. If a 25% trade discount was given by the publisher, what was the original list price?

8. Jones Tire Company got a $1,500 discount on its last purchase. If the rate of discount was 15%, what was (a) the list price and (b) the net price?

unit 7.9

Two or More Trade Discounts in a Series

After completing Unit 7.9, you will be able to:

1. Compute two or more trade discounts in a series.

In Unit 7.8, you learned that a trade discount is a percentage reduction in the list price of merchandise. You also learned that a trade discount is calculated as follows:

List price $\times$ trade discount rate = amount of trade discount

You then subtract the amount of trade discount from the list price to get the net price.

To allow for greater flexibility in pricing items, sellers sometimes offer two or more discounts in a series, for example, 20% and 10%. Such a discount is called a **chain discount** or a **series discount.** Chain discounts are especially helpful when the seller sets the discounts that are available for different classes of customers, different types of products, special promotional sales, seasonal fluctuations, and the like.

The rates in a chain discount are computed on a successive basis; that is, the first discount is calculated and subtracted; the second discount is taken on what is left. And if there is a third discount, it is taken on the amount left after the second discount is calculated and subtracted, and so on.

Example

Compute the net price of an item listed at $800 with a chain discount of 20% and 10%.

Solution

Step 1	Step 2	Step 3
$800	$800	$640
× .20	− 160	− 64
$160	$640	$576 net price
	× .10	
	$ 64	

Note that the second discount (Step 2) was calculated on the remainder after the first discount was subtracted. The second discount ($64) was then subtracted to yield a net price of $576. It does not matter what order you calculate the discounts because they are figured on a successive basis. Thus, you could have computed them as 10% and 20% (rather than as 20% and 10%) and you would get the same net price. However, *you can never add the rates together* because each discount is calculated on a successively smaller base. Let's look at another example.

Example

Compute the net price of an item listed at $1,500 with a chain discount of 20%, 10%, and 10%.

Step 1	Step 2	Step 3	Step 4
$ 1,500	$1,500	$1,200	$1,080
× .20	− 300	− 120	− 108
$300.00	$1,200	$1,080	$ 972 net price
	× .10	× .10	
	$ 120	$ 108	

We can now add the three discounts together to get the total discount: $300 + $120 + $108 = $528.

〜〜〜 EQUIVALENT SINGLE DISCOUNT

A series of discounts can be changed to a single discount that has the same value as all discounts in the series. That is, you can convert two or more discount rates in a series to an equivalent single discount rate. To illustrate, let's look again at the preceding example in which we had an item with a list price of $1,500 with a chain discount of 20%, 10%, and 10%.

To find the equivalent single discount rate:

Step 1: Subtract each discount from 100%; then find the product of these differences. (Remember that you must change percents to decimals to work with them.):

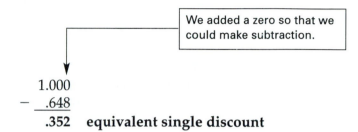

$$
\begin{array}{cccc}
1.00 & 1.00 & 1.00 \\
-\ \ .20 & -\ \ .10 & -\ \ .10 \\
\hline
.80 \ \times & .90 \ \times & .90 \ = .648
\end{array}
$$

Do not round off.

Step 2: Take the result from Step 1 (.648 in this case) and subtract it from 100%.

We added a zero so that we could make subtraction.

$$
\begin{array}{l}
1.000 \\
-\ \ .648 \\
\hline
.648 \quad \textbf{equivalent single discount}
\end{array}
$$

.352 **equivalent single discount**

Now that we found our equivalent single discount, we can multiply it times the list price to get the amount of discount:

$$
\begin{array}{ll}
\$1,500 & \text{list price} \\
\times\quad .352 & \text{equivalent single discount} \\
\hline
\$\ \ 528 & \text{amount of discount}
\end{array}
$$

Now subtract the amount of discount, $528, from the list price to get the net price:

$$
\begin{array}{ll}
\$1,500 & \\
-\quad 528 & \\
\hline
\$\ \ 972 & \text{net price}
\end{array}
$$

Note that using the equivalent single discount yielded the same net price, $972, as calculating the discounts successively, but with less work. The equivalent single discount is especially helpful when companies often buy items that have the same chain discount.

NET PRICE EQUIVALENT RATE

As you just learned, you can multiply the equivalent single discount rate times the list price of an item to get the trade discount. You then subtract the trade discount from the list price to get the net price.

We can alter this process just a bit and obtain the same results. Let's go back to our earlier example of a $1,500 item offered with a chain discount of 20%, 10%, and 10%. Remember that the first step in calculating an equivalent single discount rate is to subtract each discount from 100% (1.00) and to multiply these differences together, which we did:

$$\begin{array}{ccc} 1.00 & 1.00 & 1.00 \\ \underline{.20} & \underline{.10} & \underline{.10} \\ .80 \quad \times & .90 \quad \times & .90 \quad = .648 \end{array}$$

Now stop with this rate (instead of subtracting it from 100% as we did earlier). This rate is equivalent to the net price. So, multiplying it times the list price will yield the net price:

$$\frac{\text{List price} \times \text{net price equivalent rate}}{\$1,500 \quad \times \quad .648} = \frac{\text{net price}}{\$972}$$

To get the amount of trade discount, we can subtract the net price from the list price:

$$\begin{array}{rl} \$1,500 & \text{list price} \\ - \quad 972 & \text{net price} \\ \hline \$ \quad 528 & \text{amount of trade discount} \end{array}$$

Let's look at another example.

Example

Using the net price equivalent rate, find the net price and the amount of trade discount of an item listed at $1,200 with a chain discount of 20% and 5%.

Solution

First find the net price equivalent rate:

$$\begin{array}{cc} 1.00 & 1.00 \\ \underline{.20} & \underline{.05} \\ .80 \quad \times & .95 \quad = \textbf{.76 net price equivalent rate} \end{array}$$

Next, multiply the net price equivalent rate times the list price to get the net price:

$$\begin{array}{rl} \$1,200 & \text{list price} \\ \times \quad .76 & \text{net price equivalent rate} \\ \hline \$ \quad 912 & \text{net price} \end{array}$$

Finally, subtract the net price from the list price to find the amount of trade discount:

$$\begin{array}{rl} \$1,200 & \\ - \quad 912 & \\ \hline \$ \quad 288 & \text{amount of trade discount} \end{array}$$

Answers:

$$\begin{array}{ll} \text{Net price} & = \$912 \\ \text{Trade discount} & = \$288 \end{array}$$

Remember:

1. A chain discount is two or more discounts in a series.
2. You never add discounts in a series together because the discounts are taken on successively smaller bases.
3. Multiplying the list price by the equivalent single discount rate yields the trade discount.
4. Multiplying the list price by the net price equivalent rate yields the net price.

Using the net price equivalent rate, find (a) the net price and (b) the amount of trade discount of an item listed at $1,800 with discounts of 20% and 10%.

Solution

$$\begin{array}{cc} 1.00 & 1.00 \\ -\ .20 & -\ .10 \\ \hline .80\ \times & .90 \end{array}\ =\ .72\ \text{net price equivalent rate}$$

(a) $1,800 \times .72 = $1,296 net price
(b) $1,800 - 1,296 = $504 trade discount

Work Space

Exercises

Find the equivalent single discount rate for each of the following. Round off decimals to 4 places where necessary.

1. 30%, 20%, and 10%

2. 20% and 10%

3. 20% and 20%

4. 20%, 25%, 5%, and 10%

5. 20%, 20%, and 10%

6. 10%, 10%, and 10%

7. 5%, 10%, and 5%

8. $33\frac{1}{3}$%, 15%, and 5%

9. $7\frac{1}{2}$%, 10%, and 10%

10. 40%, 20%, and 5%

11. 10% and 10%

12. 5% and $2\frac{1}{2}$%

ANSWERS

1. _____
2. _____
3. _____
4. _____
5. _____
6. _____
7. _____
8. _____
9. _____
10. _____
11. _____
12. _____

Using equivalent single discount rates, find the amount of trade discount and the net price for each of the following. Round to 4 places where necessary, and round all dollar amounts to even cents:

	List Price	Chain Discount	Amount of Trade Discount	Net Price
13.	$ 800	20% and 12%	$_____	$_____
14.	$ 900	$33\frac{1}{3}$% and 15%	_____	_____
15.	$ 1,450	25%, 15%, and 5%	_____	_____
16.	$ 2,500	40% and 10%	_____	_____
17.	$ 6,400	10%, 5%, and 2%	_____	_____
18.	$ 9,800	20%, 20%, and 10%	_____	_____
19.	$12,500	25%, 20%, and 5%	_____	_____
20.	$60,450	15%, 12%, and 5%	_____	_____

Work Space

Applications

In the following problems, round all rates where necessary to 4 places, and round all dollar amounts to even cents:

ANSWERS

1. Find the net price of a walk-in cooler if the cooler has a list price of $9,000 and is offered with a chain discount of $33\frac{1}{3}\%$, 20%, and 10%.

1. _____

2. _____

3. a. _____

   b. _____

2. Tom Jarrel is thinking about buying a freezer. He found two units that he liked. One cost $1,200 with a chain discount of 20%, 10%, and 5%. The other cost $1,255 with a chain discount of 20%, 12%, and 4%. Which should he buy? Why?

3. Edge Manufacturing Company sold a set of tools to American Buildings Company. The list price of the tools was $1,400 with a chain discount of 10%, 10%, and 2%. What was (a) the amount of trade discount and (b) the net price?

4. Lee King bought a welder and got a $171 trade discount. If the welder was offered a chain discount of 10% and 10%, what was the list price? (*Hint:* $171 = P$; 10% and 10% = R; find B.)

5. An item with a list price of $1,500 was bought for $1,200. What rate of trade discount was received?

6. What is the amount of trade discount and the net price of an item listed at $12,000 with a chain discount of $33\frac{1}{3}$%, 20%, and 10%? (*Helpful Hint:* Use aliquot parts instead of decimals.)

7. Which of the following chain discounts will yield the greater discount: (a) 20%, 20%, and 10% or (b) 20%, 18%, and 12%? How much more? (*Hint:* Convert the discounts to equivalent single discount rates and compare the single rates.)

8. Differentiate between the *equivalent single discount rate* and the *net price equivalent rate*.

unit 7.10

Cash Discounts

After completing Unit 7.10, you will be able to:

1. Compute cash discounts.

In Units 7.8 and 7.9, we discussed trade discounts, which, as you recall, are deductions from a catalog or list price. Another type of discount often given by sellers is the **cash discount,** which is a discount offered to encourage customers to make prompt and early payment. The rate of a cash discount and the time the customer has to take advantage of the cash discount are part of the **credit terms** that will appear on the invoice.

A common form of cash discount is **2/10,n/30** (read as **two ten, net thirty**). This means that a 2% discount can be taken if the invoice is paid within 10 days from the invoice date; if payment is not made within 10 days, the full amount must be paid within 30 days from the invoice date. If payment is not made within 30 days, the debt is past due and the seller can charge interest or a late fee. Let's look at an example of an invoice that is paid within the discount period.

Example

An item with an invoice price of $800 was purchased on June 2 with terms of 2/10, n/30. If payment was made on June 12, what was the net amount paid?

Solution

Step 1: Compute the cash discount:

$$
\begin{array}{ll}
\$800 & \text{invoice price} \\
\times \ \underline{\ .02} & \\
\$\ 16 & \text{cash discount}
\end{array}
$$

Step 2: Subtract the cash discount from the invoice price:

$$
\begin{array}{r}
\$800 \\
-\ \underline{16} \\
\$784 \quad \text{net amount paid}
\end{array}
$$

Cash discounts are only taken on the price of merchandise; they are *not* taken on charges for such things as freight and packing. If charges for freight and packing are included in the invoice, they must be deducted before the cash discount is taken. They are then added back in order to get the total amount due.

Example

Phillips Company received goods that were invoiced for $780. The invoice price included a freight charge of $30. What is the amount due if the invoice carried terms of 2/10, n/30 and payment was made within the discount period?

Solution:

Step 1: Subtract the freight charge from the invoice price:

$$
\begin{array}{rl}
\$780 & \text{total of invoice} \\
-\ \underline{30} & \text{freight charge} \\
\$750 & \text{amount subject to cash discount}
\end{array}
$$

Step 2: Multiply the rate of cash discount times the amount subject to the discount:

$$
\begin{array}{rl}
\$750 & \\
\times\ \underline{.02} & \text{rate of cash discount} \\
\$\ 15 & \text{amount of cash discount}
\end{array}
$$

Step 3: Subtract the cash discount and add back the freight:

$$
\begin{array}{rl}
\$750 & \\
-\ \underline{15} & \\
\$735 & \\
+\ \underline{30} & \text{freight charge} \\
\mathbf{\$765} & \text{total amount due}
\end{array}
$$

 OTHER COMMON TERMS OF SALE

Terms of 2/10, n/30 are very common. But there are other common terms of sale. Look at the following chart to see some of them.

Now, before starting the exercises, let's look at a couple of other examples.

Terms	Read as	Meaning
n/30	"net thirty"	No discount. Payment is due within 30 days from the invoice date.
c.o.d.	"cash on delivery"	Payment is due when the merchandise is delivered to the buyer.
2/10,n/e.o.m.	"two ten, net end of month"	A 2% discount can be taken if payment is made by the 10th day of the month that *follows* the month of sale. If the discount is not taken, payment must be made within 20 days after the discount period.
2/10,n/30 r.o.g.	"two ten, net thirty receipt of goods"	A 2% discount can be taken if payment is made within 10 days **after** receipt of the goods; full amount is due between day 11 and day 30 if payment is not made within discount period.

Example 1

On May 18, Rodriguez Company bought goods with an invoice price of $900. The invoice carried terms of 2/10, n/e.o.m. and payment was made on June 9. What is the net amount due?

Solution

The discount applies because payment was made within 10 days after the end of the month:

$$\$900 \times .02 = \$18; \$900 - \$18 = \textbf{\$882}$$

Example 2

Taylor Supply Company bought $6,000 worth of merchandise on July 6. The merchandise was delivered on July 11 and payment was made on July 21. What is the net amount paid if terms of sale were 2/10, n/30 r.o.g.?

Solution

The discount applies because payment was made on July 21, which is 10 days after the merchandise was received:

$$\$6,000 \times .02 = \$120; \$6,000 - \$120 = \textbf{\$5,880}$$

Remember:

1. A cash discount is offered to encourage prompt payment of an invoice; it is not automatic.
2. The cash discount period starts running with the date of the invoice, not the actual date of delivery.
3. Only the actual price of goods is subject to a cash discount. And even though they must be added back after the cash discount is taken, additional charges for such things as freight and packing are not subject to the discount.

Work Space

Name _____ Date _____

Exercises

Find the amount of cash discount and the net amount paid for each of the
following. Round all dollar amounts to even cents:

	Invoice Amount	Invoice Date	Terms	Date of Payment	Cash Discount	Net Amount Paid
1.	$ 1,800	Jan. 4	2/10,n/30	Jan. 14	$_____	$_____
2.	4,500	Mar. 12	1/10,n/30	Mar. 18	_____	_____
3.	600	Oct. 28	3/20,n/90	Nov. 15	_____	_____
4.	1,250	June 30	3/10,2/20,n/45	July 20	_____	_____
5.	12,500	Aug. 12	4/10,n/30	Aug. 22	_____	_____
6.	62,578	Dec. 1	2/10,1/30,n/90	Dec. 31	_____	_____
7.	22,400	Feb. 3	2/10,1/20,n/30	Mar. 5	_____	_____
8.	34,800	Mar. 7	2/10,n/e.o.m.	Apr. 9	_____	_____
9.	12,800	June 14	1/10,n/90	June 24	_____	_____
10.	45,675	Oct. 6	2/10,n/e.o.m.	Nov. 10	_____	_____

Work Space

Applications

1. Goods with an invoice price of $525 were bought on June 18. The invoice carried terms of 2/10,n/30 and payment was made on June 28. What is (a) the amount of cash discount and (b) the net amount paid?

ANSWERS

1. a. _____

   b. _____

2. a. _____

   b. _____

3. _____

4. _____

2. On March 2, Carlsbad Company bought $3,000 worth of merchandise from Lincoln Company. Terms of sale were 3/10,2/20,n/45. What was the net amount paid if Carlsbad made payment on (a) March 22 and (b) March 11?

3. On July 14, Simmers Company received goods invoiced at $2,740 from Tyler Company. The invoice was dated July 12 and included a freight charge of $35. If terms of payment were 2/10,n/30, and payment was made on July 23, what was the net amount paid?

4. Mills, Inc. received a shipment on October 15. The invoice, dated October 14, carried a list price of $1,800 subject to a 10% trade discount and terms of 2/10,n/30. Also shown on the invoice was a freight charge of $25 and a packing charge of $10. If Mills made payment on October 24, what was the net amount paid?

5. Find the net amount to be paid for each of the following invoices:

	Invoice Date	List Price	Trade Discount	Terms	Date of Payment	Net Amount Paid
a.	Jan 15	$4,000	none	2/10,n/30	Jan. 25	$_____
b.	Apr. 10	5,600	10% and 10%	3/10,2/15,n/30	Apr. 24	_____
c.	May 28	7,200	20%, 10%, and 5%	2/15,n/90	July 22	_____
d.	Feb. 4	9,000	$33\frac{1}{3}\%$	3/20,n/45	Feb. 24	_____

6. Goods with an invoice price of $7,400 were received on November 12. The date of the invoice was November 10 and terms of sale were 3/10,2/20,1/30,n/90. What was the net amount paid if payment was made on:

   a. November 22

   b. November 20

   c. November 30

   d. December 1

   e. December 31

   f. January 15 (of the following year)

7. A drill press with an invoice price of $34,800 was received on June 5. The date of the invoice was June 2 and terms of sale were 2/10,n/30 r.o.g. What was the net amount paid if payment was made on (a) June 15 and (b) June 22?

8. Goods invoiced at $800 were received on July 14. Terms of sale were 2/10,n/e.o.m. What was the net amount paid if payment for the goods was made on August 10?

6. a. _____

   b. _____

   c. _____

   d. _____

   e. _____

   f. _____

7. a. _____

   b. _____

8. _____

Appendix

Working with Measurements

Some business math problems will require that you use various types of measurements. Measures of length, weight, surface, liquid, volume, time, and dozens are often used. The following table shows the measures that are most commonly used in business.

     Problems involving measures generally are of three types:

1. Converting from one unit of measure to another
2. Adding, subtracting, multiplying, or dividing measures
3. Solving problems through the application of measures

≈ CONVERTING FROM ONE UNIT OF MEASURE TO ANOTHER

Use Table A–1 on page 522 to help you when converting from one unit of measure to another. To convert a smaller unit of measure into a larger unit, you must divide.

Example 1
How many feet are in 168 inches?

Solution

$$12 \text{ in.} = 1 \text{ ft.}$$
$$168 \div 12 = 14 \text{ feet}$$

Example 2
A large steak weighs 56 ounces. What is the weight of the steak in pounds?

Solution

$$16 \text{ oz.} = 1 \text{ lb.}$$
$$56 \div 16 = 3.5 \text{ lbs. (or 3 lbs. or 8 oz.)}$$

To convert a larger unit of measure to a smaller unit, you need to multiply.

Table A–1

Measurements	
Weight	
16 ounces (oz.)	= 1 pound (lb.)
2,000 pounds	= 1 ton
Length (Linear)	
12 inches (in.)	= 1 foot (ft.)
3 feet	= 1 yard (yd.)
5,280 feet	= 1 mile (mi.)
1,760 yards	= 1 mile
Time	
60 seconds (sec.)	= 1 minute (min.)
60 minutes	= 1 hour (hr.)
24 hours	= 1 day (da.)
7 days	= 1 week (wk.)
$4\frac{1}{3}$ weeks	= 1 month (mo.)
52 weeks	= 1 year
12 months	= 1 year
Capacity	
2 cups	= 1 pint (pt.)
2 pints	= 1 quart (qt.)
4 quarts	= 1 gallon (gal.)
16 ounces (1 lb.)	= 1 pint
Area (Surface)	
144 square inches	= 1 square foot (sq. ft.)
9 square feet	= 1 square yard (sq. yd.)
43,560 square feet	= 1 acre (a.)
640 acres	= 1 square mile (sq. mi.)
Volume (Cubic)	
1,728 cubic inches	= 1 cubic foot (cu. ft.)
27 cubic feet	= 1 cubic yard (cu. yd.)
1 cubic foot	= $7\frac{1}{2}$ gallons of water
Other Measures	
12 units	= 1 dozen (doz.)
12 dozen	= 1 gross (gr.)
144 units	= 1 gross

Example 1

How many ounces are in $6\frac{1}{4}$ pounds?

Solution

$$1 \text{ lb.} = 16 \text{ oz.}$$

$$6\frac{1}{4} = 16 = \frac{25}{\cancel{4}_1} \times \frac{\cancel{16}^4}{1} = 100 \text{ oz.}$$

Example 2

Ben Taylor weighs 190 lbs. What is his weight in ounces?

Solution

$$
\begin{array}{rl}
190 & \text{lbs.} \\
\times \quad 16 & \text{ounces in 1 lb.} \\
\hline
3{,}040 & \text{ounces in 190 lbs.}
\end{array}
$$

Example 3

How many quarts are in 12 gallons?

Solution

$$1 \text{ gal.} = 4 \text{ qt.}$$
$$12 \text{ gal.} \times 4 = 48 \text{ qt.}$$

ADDING, SUBTRACTING, MULTIPLYING, AND DIVIDING MEASURES

Addition

To add measures, you add the similar units separately and then convert the excess units.

Example 1

Add: 6 yards, 1 foot, 3 inches + 4 yards, 2 feet, 7 inches.

Solution

Yards	Feet	Inches
6	1	3
4	2	7
10	3	10
+ 1	−3	
11		10

Add similar units together.

Convert excess units: 3 ft. = 1 yd.

Example 2

Add: 7 yards, 1 foot, 10 inches + 8 yards, 2 feet, 9 inches.

Solution

Yards	Feet	Inches
7	1	10
8	2	9
15	3	19
	+1	−12
15	4	7
+ 1	−3	
16	1	7

10 ← Add similar units.

−12 ← Convert excess in. to ft.

−3 ← Convert excess ft. to yd.

Subtraction

When subtracting measures, as in addition, you subtract similar units separately.

Example

Subtract: 3 hours, 15 minutes from 5 hours, 30 minutes.

Solution

Hours	Minutes
5	30
−3	15
2	15

In some subtraction problems, you will have to borrow from the larger unit.

Example

Subtract: 12 yards, 2 feet, 7 inches − 4 yards, 11 inches.

Solution

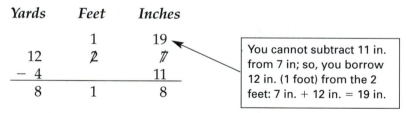

Yards	Feet	Inches
	1	19
12	2̶	7̶
− 4		11
8	1	8

You cannot subtract 11 in. from 7 in; so, you borrow 12 in. (1 foot) from the 2 feet: 7 in. + 12 in. = 19 in.

Multiplication

To multiply units of measure, you follow these steps:

Step 1: Convert the units being multiplied to one common unit.

Step 2: Multiply as usual.

Example

Multiply 7 ft. 6 in. by 5 ft. 4 in.

Solution

Step 1: Convert to one common unit:

$$7 \text{ ft. } 6 \text{ in. } = 7\frac{6}{12} \text{ ft. } = 7\frac{1}{2} \text{ ft.}$$
$$5 \text{ ft. } 4 \text{ in. } = 5\frac{4}{12} \text{ ft. } = 5\frac{1}{3} \text{ ft.}$$

> 12 in. = 1 ft.; thus, 6 in. = $\frac{6}{12}$ ft.

Step 2: Multiply:

$$7\frac{1}{2} \times 5\frac{1}{3} = \frac{\overset{5}{\cancel{15}}}{\underset{1}{\cancel{2}}} \times \frac{\overset{8}{\cancel{16}}}{\underset{1}{\cancel{3}}} = 40 \text{ sq. ft.}$$

Division

To divide units of measure, you convert the units to one common unit and divide in the normal way.

Example

Divide 6 gal. 3 qt. by 3.

Solution

Convert to one common unit:

$$6 \text{ gal. } 3 \text{ qt. } = 6\frac{3}{4} \text{ gal.}$$

> 1 qt. = $\frac{1}{4}$ gal.; thus, 3 qt. = $\frac{3}{4}$ gal.

Divide:

$$6\frac{3}{4} \div 3 = \frac{\overset{9}{\cancel{27}}}{4} \times \frac{1}{\underset{1}{\cancel{3}}} = \frac{9}{4} = 2\frac{1}{4} \text{ gal.}$$

SOLVING BUSINESS PROBLEMS INVOLVING MEASUREMENTS

Many everyday business problems call for the use of measurements. For example, the cost of carpet is quoted in terms of square yards. Let's assume you wanted to carpet your den with carpet that is advertised at "$12 per square yard, installed." Carpeting covers an area (or a surface). Thus, you will be working with square measures. The first thing you will need to do is to measure the length and width of the room. Let's assume the room is 24 feet long and 18 feet wide. Now you obtain the square feet in your den by multiplying length by width:

$$24 \times 18 = 432 \text{ sq. ft.}$$

Now look at Table A–1 and you find that there are 9 square feet to the square yard. So, you divide 432 square feet by 9 to obtain the number of square yards:

$$432 \div 9 = 48 \text{ sq. yds.}$$

You now find the price of the carpet by multiplying the cost of the carpet by the number of square yards needed:

$$48 \text{ sq. yds.} \times \$12 \text{ per sq. yd.} = \$576$$

Let's look at another example.

Example

How many gallons will it take to fill a tank that is 20 feet long, 7 feet wide, and 4 feet high?

Solution

$$\frac{\text{Length}}{20} \times \frac{\text{width}}{7} \times \frac{\text{height}}{4} = \frac{\text{volume}}{560 \text{ cu. ft.}}$$

$$560 \text{ cu. ft.} \times 7\frac{1}{2} \text{ gal. per cu. ft.} = 4{,}200 \text{ gal.}$$

Work Space

Appendix Exercises

1. How many feet are there in:

    a. 228 inches _____

    b. 276 inches _____

    c. 78 inches _____

2. How many ounces are there in:

    a. 8 pounds _____

    b. 16 pounds _____

    c. $5\frac{3}{4}$ pounds_____

3. How many pints are in 12 gallons?

4. How many hours are in 5,400 seconds?

ANSWERS

3. _____
4. _____

5. Convert the following:

  a. 54 inches to feet _____

  b. 36 feet to yards_____

  c. 5,184 cubic inches to cubic feet _____

  d. 63 square feet to square yards _____

  e. 3 miles to yards _____

  f. 128 ounces to pounds _____

  g. 21,120 feet to miles _____

  h. 8 tons to pounds _____

  i. 3 acres to square feet_____

6. Add the following:

  a. 6 yards, 2 feet, 7 inches + 5 yards, 1 foot, 11 inches

  b. 12 yards, 9 inches + 3 yards, 2 feet, 10 inches + 8 yards, 1 foot

  c. 6 hours, 45 minutes + 3 hours 15 minutes

  d. 4 hours, 20 minutes, 30 seconds + 2 hours, 40 minutes, 50 seconds

ANSWERS

6. a. _____

b. _____

c. _____

d. _____

7. Subtract the following:

a. 6 hours, 18 minutes − 3 hours, 30 minutes

b. 12 hours, 12 minutes, 48 seconds − 4 hours, 10 minutes, 52 seconds.

c. 5 yards, 2 feet, 3 inches − 2 yards 1 foot, 8 inches

8. Perform the following:

ANSWERS

8. a. _____

b. _____

c. _____

d. _____

a. Multiply 3 feet, 4 inches by 2 feet, 6 inches

b. Multiply 2 feet, 6 inches by 10 inches

c. Divide 4 gallons, 2 quarts by 2

d. Divide 6 feet, 8 inches by 5

Name _____ **Date** _____

Applications

1. How many square yards of linoleum tile would it take to cover a kitchen that is 30 ft. wide and 36 ft. long?

2. If a 6 ft. 9 in. pipe is cut into three equal pieces, what would be the length of each piece?

3. How many gallons of water would it take to fill completely a tank that is 24 ft. by 8 ft. by 6 ft.?

4. Sue Taylor weighs 1,680 ounces. What is her weight in pounds?

5. A room 24 ft. by 18 ft. is to be carpeted. How much would it cost to cover the room if the carpeting is $8.00 per square yard, installation is $2.00 per square yard, and padding is $1.50 per square yard?

6. Alan can run a mile in 8 minutes and 18 seconds. Todd can run a mile in half Alan's time. What is Todd's time?

ANSWERS

1. _____

2. _____

3. _____

4. _____

5. _____

6. _____

Work Space

Index